Economics of Offshore Wind Power

Rahmatallah Poudineh • Craig Brown
Benjamin Foley

Economics of Offshore Wind Power

Challenges and Policy Considerations

Rahmatallah Poudineh
The Oxford Institute for Energy Studies
Oxford, UK

Craig Brown
The Oxford Institute for Energy Studies
Oxford, UK

Benjamin Foley
Keystone Engineering Inc.
New Orleans, LA, USA

ISBN 978-3-319-66419-4   ISBN 978-3-319-66420-0   (eBook)
https://doi.org/10.1007/978-3-319-66420-0

Library of Congress Control Number: 2017957581

Cover illustration: Modern building window © saulgranda/Getty

Printed on acid-free paper

This Palgrave Macmillan imprint is published by Springer Nature
The registered company is Springer International Publishing AG
The registered company address is: Gewerbestrasse 11, 6330 Cham, Switzerland

FOREWORD

Continuous advancements in wind generation technology coupled with falling costs and increased interest in moving towards a low-carbon energy mix suggest a bright future for wind—and solar—energy. In recent competitive auctions, wind and solar have beaten conventional generation options with prices that continue to decline. This suggests that subsidies and other financial incentives, still offered in many parts of the world, including the United States, may be phased out over time as the industry becomes competitive.

In a recent competitive auction for offshore wind in Germany, for example, 1300 MW out of 1450 were accepted without *any* subsidies. Until recently, bidders would have received a 12 €cts/kWh subsidy for every kWh generated over 20 years. The offshore wind also achieved a low clearing price in the UK Contracts for Difference (CfD) second allocation round (£57.50/MWh). These, and similar trends, demonstrate that wind—and solar—technologies can indeed achieve autonomy as the industries evolve under the right circumstances.

While onshore wind has enjoyed rapid growth in key markets around the globe, many experts, including the authors of this volume, see major opportunities for offshore wind moving forward. Despite higher installation and maintenance costs, offshore wind offers lower generation costs due to more persistent and higher-velocity winds. Moreover, offshore wind farms can be positioned out of sight and out of mind in order to lower the visual and environmental opposition.

The variability of renewables—which applies to both wind and solar—has become more pronounced in places with high penetration of these resources such as Europe, the USA, China, Australia, and elsewhere. This clearly requires more investment in transmission lines, better interconnections, more storage, and certainly more demand participation in the market—all areas where investments are expected to take place.

And this is where regulators and policy makers can—and should—play a constructive role by encouraging further investments in low-carbon energy resources while providing the means to integrate their variable output into existing grids, historically designed to deliver power from large, central, dispatchable thermal stations to major load centres.

As we move forward into a future where the predominant form of generation will increasingly come from variable renewable resources, policy makers and regulations must provide not only certainty, but the incentive to invest in modernising the transmission and delivery infrastructure. This includes more cost-reflective tariffs that potentially vary by time and location to offer incentives to consumers to shift their consumption patterns in response to variable wholesale prices.

Moreover, for many experts, such as the authors of this volume, it is becoming clear that the traditional industry paradigm—where the output of various types of thermal plants was adjusted to follow the variations in demand—must be replaced by a new paradigm—where demand takes an increasingly active role in response to variable prices in the wholesale markets due to swings in variable generation. Only when *price-responsive demand* learns to tango with *variable renewable generation* can the grid of the future accommodate large swings in renewable generation, increasingly becoming the norm as the world moves towards low-carbon energy sources.

The contribution of the authors of this book is noteworthy since they focus on the promise of offshore wind, which offers tremendous growth opportunities in many parts of the world where significant resources remain untapped.

Needless to say, policy makers need to focus on incentivising cost reduction in the offshore wind industry by implementing appropriate support policies such as competitive auctions, creating a supply chain, reducing the risks, encouraging technological innovation, investing in Research and Development, and providing a stable environment for project financing.

Menlo Energy Economics,   Fereidoon P. Sioshansi
San Francisco, CA, USA
June 2017

PREFACE

In 2015, we authored a working research paper on policies to promote the offshore wind power industry. The work, which was published as a paper in the Oxford Institute for Energy Studies, became a source of inspiration for us to look deeper into this industry and, in particular, the implications of our original findings as the industry expanded globally. The opportunity to write this book came about when Anna Reeve from Palgrave Macmillan approached us to write an energy-related book. Given our expertise and interests in the offshore wind industry, along with the fact that, at that time, there was not a single publication that offered a broad but concise view of the industry economics and its challenges, we decided to write this book.

The book is composed of ten chapters starting with an introduction that discusses the history and main drivers of offshore wind power, along with its role in achieving wider government policies. It then goes on to present an overview of the global offshore wind market in Chap. 2, along with current market structure and latest developments. The key challenges of the industry, which are cost drivers and technical hurdles, are discussed in Chap. 3. The book then presents approaches to support renewable electricity, including offshore wind, in Chap. 4, with a view to the necessity of harmony between support policy and wider electricity industry and market characteristics. The current support schemes applied in various offshore wind markets around the world are analysed in Chap. 5, before Chap. 6 provides a detailed analysis of cost reduction opportunities and innovation in the industry. Public acceptance, which is often an overlooked subject when it comes to offshore wind, is discussed in Chap. 7. In Chap. 8, we review the current trajectory of offshore wind in the global market and highlight the key uncertainties that might affect this trend over the next 5–10 years. In Chap. 9, we present the main policy considerations in order to create a vibrant offshore wind market that has a prospect of becoming a competitive industry. Finally, in Chap. 10, we provide the concluding remarks.

The purpose of this book is to be an informative source about the offshore wind industry for students, academics, policy makers, industry experts, and interested readers by providing a short but broad and widely accessible analysis of the industry. By reading this book, the reader can expect to learn how the offshore wind industry has been developed to date in key markets, why it has been sluggish compared with its onshore counterpart despite nearly 30 years of history, what are the challenges facing this industry, what is the role of government policies in realising cost reductions, and how policies can be formulated to nurture a sustainable industry. The book can be part of the reading list in a course in renewable energy or wind energy, or alternatively a course in wider energy economics. Although the book is written in a way that it offers a coherent story from the beginning to the end, readers can directly refer to chapters they are interested in without the need of reading the whole book. Along with being a part of the bigger picture, each chapter provides a self-contained analysis of its topic.

We would like to thank our friends, colleagues, and especially our families for their support and encouragement during this journey. During the year, while we were busy with this project, we sometimes had to dedicate some of our family times to carry out research and to draft this manuscript. Thus, we are truly grateful for the support we received from our families and hopeful that the contribution that this book makes to the debate about offshore wind economics can make up for long hours and lost time. Also, we would like to extend our gratitude to Anna Reeve and Thomas Coughlan from Palgrave Macmillan, who supported publication of our work. Lastly, we would be remiss if we did not acknowledge and thank the various professionals and colleagues working tirelessly in offshore wind engineering, policy design, and project development to bring this technology to the forefront of global clean energy markets.

Oxford Institute for Energy Studies   Rahmatallah Poudineh
Oxford, UK

Oxford Institute for Energy Studies   Craig Brown
Oxford, UK

Keystone Engineering Inc.   Benjamin Foley
New Orleans, LA, USA

CONTENTS

LIST OF ABBREVIATIONS

AEP   Annual energy production
CAPEX   Capital expenditure
CfD   Contract for difference
DONG   Danish Oil and Natural Gas
EEG   German Renewable Energy Act
EPCI   Engineering, procurement, construction, and installation
ESP   Electrical services platform
ETS   Emission trading scheme
EU   European Union
FID   Final investment decision
FiP   Feed in premium
FiT   Feed in tariff
GHG   Greenhouse gas
HVAC   High-voltage alternating current
HVDC   High-voltage direct current
ITC   Investment tax credit
JV   Joint venture
LCOE   Levelised cost of energy
MENA   Middle East and North Africa
NETSO   National Electricity Transmission System Operator
NIMBY   Not in my backyard
NREAP   National Renewable Energy Action Plan
NSCOGI   North Sea Countries Offshore Grid Initiative
OEM   Original equipment manufacturer
OFTO   Offshore transmission owner
OPEX   Operating expenditure
OREC   Offshore Renewable Energy Credit

O&M	Operation and maintenance
PPA	Power purchase agreement
PTC	Production tax credit
PURPA	Public Utilities Regulatory Policy Act
REC	Renewable Energy Certificate
RO	Renewable obligation
ROC	Renewable Obligation Certificate
RPS	Renewable Portfolio Standard
R&D	Research and Development
TGC	Tradable Green Certificate
TIV	Transportation and installation vessels
TSO	Transmission system operator
WTG	Wind turbine generator

LIST OF FIGURES

List of Tables

CHAPTER 1

Background: Role of the Offshore Wind Industry

Abstract The offshore wind industry has evolved both as an extension of and as an alternative to land-based wind farms in countries hindered by land scarcity or constrained by public concerns. The social and political role that offshore wind plays is similar to that of land-based wind and other renewable energy technologies, namely, that offshore wind farms bolster energy security, support greenhouse gas emission reductions and renewable energy targets, and create opportunities for new industries and jobs. However, the unique attributes of offshore wind farms, such as the greater wind energy resources they are able to capture offshore and their proximity to major coastal demand centres, have also helped to carve out a unique political, economic, and public role for the offshore wind industry.

Keywords Renewable energy targets • Kyoto Protocol • European energy package • Paris Agreement • Role of offshore wind

1.1   Introduction

The role that renewable energy plays in a country is heavily intertwined with the political, geopolitical, and social context influencing policy makers and their constituents. The Arab Oil Embargo of the 1970s altered the way in which governments responded to energy security issues, compelling many developed nations including the United States and United

© The Author(s) 2017
R. Poudineh et al., *Economics of Offshore Wind Power*,
https://doi.org/10.1007/978-3-319-66420-0_1

Kingdom to direct public funds into the Research and Development (R&D) of new energy technologies, including wind energy, to bolster energy security of supply. The Peak Oil threat, in full force by the late 1970s, furthermore helped to buoy political and social support for alternative energy technologies in the last decades of the twentieth century.

By the turn of the twenty-first century, the political rationale for renewables began to expand from solely energy security of supply to climate change mitigation. The Kyoto Protocol, adopted in December 1997 and implemented in 2005, created the first binding international treaty for governments to reduce greenhouse gas (GHG) emissions. This gave rise to increasing government support for the R&D and commercialisation of new renewable energy technologies, particularly within the European Union (EU). The Paris Agreement of 2015 furthered these efforts by achieving widespread consensus for the mitigation of GHG emissions, largely through targeting decarbonisation of power generation sectors. These approaches were based on the strategy that decarbonising the electricity sector paves the way for further decarbonisation of economies, for example, through electrification of the heat and transport sectors.

Offshore wind has carved out a distinctive role within this evolving narrative of supporting renewable energy technologies. Having originated in northern Europe over 25 years ago, the offshore wind industry emerged, primarily, as an incipient industry to the land-based wind industry in the early 1990s in European countries, where land scarcity and land-use issues impeded the potential for an onshore wind industry to proliferate. The unique features of offshore wind farms, notably the stronger wind energy resources they could capture offshore and their proximity to the major coastal demand centres, moreover defined a particularly unique role for the technology. By 2010, offshore wind capacity targets were appearing prominently in individual EU members' National Energy Action Plans. As the industry has taken shape in Europe, several other countries have also started to explore the feasibility of offshore wind over the past decade, including the United States, India, China, Taiwan, and Vietnam. Today, offshore wind energy is on the cusp of becoming a mainstream renewable power generation source.

This chapter reviews the background and the role of the offshore wind sector in achieving countries' energy policy objectives. The next section presents the historical trajectory leading to the emergence of offshore wind power in the renewable energy mix. Sections 1.3 and 1.4 discuss the key role that offshore wind energy fulfils in the current social and political context within Europe and outside Europe, respectively. Finally, Sect. 1.5 provides concluding remarks.

1.2   HISTORY

Humankind has utilised wind energy since the dawn of civilisation. The first accounts of wind being used to propel ships across bodies of water date back to the fifth millennium BCE in Mesopotamia. Throughout the seventh to ninth century AD, vertical axis panemone windmills were utilised in Persia for practical purposes such as grinding flour and pumping water (Shahan 2014). By the twelfth century, horizontal axis windmills started to appear in northern Europe, where they were used as grinding mills for grains. Figure 1.1 depicts the remains of a historical windmill in the Sistan region of Iran.

The first inclinations of using wind to produce electricity emerged during the 1880s. In 1882, the first of Thomas Edison's coal-fired electric-light stations was built in London's Holborn Viaduct Station and followed shortly thereafter by Manhattan's Pearl Station, providing DC electrical power supply to nearby customers. By 1887, over 120 Edison power stations had sprouted up in the United States (Swift-Hook 2012). The concept of using wind power to drive the rotating electrical dynamo had also first emerged around this time, in the early 1880s. Accounts of the first generation of electricity from a wind-driven machine vary, but the invention is widely credited to Prof. James Blyth of Anderson's College in Scotland (now University of Strathclyde) in July 1887 (Swift-Hook 2012). Blyth used electricity from a wind-powered generator to power his vacation cottage at Marykirk, Scotland, effectively making it the first home in the world to be supplied electricity by wind energy.[1]

By 1895, the Danish scientist and inventor Poul la Cour had further improved upon the wind turbine technology. It is interesting to note that la Cour's zeal for the technology was driven in part by a social agenda, as well as the science and innovation. Having been raised on a farm, la Cour adamantly sought to utilise the technology to strengthen the appeal of rural areas of Denmark that were facing rapid abandonment in the frenzy of European industrialisation (Stenkjaer 2009). By the early twentieth century, land-based wind turbines were being used across Denmark primarily to power rural sites, including small homes, schools, farms, and villages. By 1908, some 72 wind turbine generators had been installed across the country (Shahan 2014).

Fig. 1.1 The remains of a historical windmill in the Sistan region, Iran (Source of image: Taken by authors)

Although wind turbines were being increasingly adopted, the commercial industry wouldn't begin to evolve until nearly two decades later. In the late 1920s, industrialists in the United States first began commercialising small wind turbine generators for agricultural purposes. By the 1930s, roughly 30,000 small wind turbines were in use across the United States (Rivkin et al. 2014).

Offshore wind would evolve both as an extension of and an alternative to land-based wind turbines. The first documented theoretical approaches for installing wind turbines offshore had originated in the early 1930s in Germany (Kaldellis and Kapsali 2013). Nevertheless, a series of practical and technological impediments of the time prohibited their application. Amongst a multitude of other constraints, there was no offshore industry to install the turbines, let alone subsea cable technology to export power to land. As such, for decades to come, offshore wind would remain merely conceptual rather than a realistic endeavour.

It was the Arab Oil Embargo of 1973 that would ultimately spur the development of the necessary technologies that the offshore wind industry could eventually grow from. Seeking to reduce the vulnerability of the United States to oil price shocks and supply disruption, the US Department of Energy and the Department of Interior mandated NASA to establish a programme to research and develop utility-scale wind turbines in 1975. The NASA programme proved to be invaluable in pioneering the large, multi-megawatt (utility-scale) turbine technologies, developing many of the composite blade materials, variable speed generators, steel tube towers, and various aerodynamic and structural engineering capabilities and technologies that would later be adopted by the industry (Shahan 2014). This would also be critical in moving wind energy generation from small, rural applications to mainstream utility power generation.

Concomitant to this, the US federal Public Utilities Regulatory Policy Act of 1978 (PURPA) had also first mandated that utilities interconnect renewable energy projects to the grid. The first land-based wind farm in the world was erected shortly thereafter in 1980 in New Hampshire, United States, and consisted of twenty 30 kW turbines (600 kW total capacity). Even though this first commercial wind farm was neither a technical nor an economic success (the turbines broke down and the developers overestimated wind resources), entrepreneurship and policy incentives managed to keep the wind power industry alive. In 1981, the state of California first introduced tax credits and rebates for wind turbines, ushering in an exponential growth of the industry over just a few short years.

In the early 1990s, the Federal Solar, Wind, Waste and Geothermal Power Production Incentives Act then removed size limitations on renewable energy plants qualifying for PURPA benefits, and in 1992, the US federal government implemented the production tax credit (PTC) for wind power for the first time. Notably, the PTC incentivised electricity production rather than just installation, which effectively encouraged improvements in performance, reliability, and technology efficiency across the industry (Shahan 2014). These developments drove the first 'boom' cycle for the wind industry in the United States, which saw not only capacity gains but also technological improvements. By the year 2000, there were nearly 100 utility-scale wind farms operating across the United States with an installed capacity of approximately 2500 MW.

1.2.1   Emergence of Offshore Wind

Like the United States, the UK government had also been compelled by the Oil crisis of 1973 to consider the development of wind power and had started to implement R&D programs for wind energy by the mid-1970s (Kern et al. 2014). However, commercial wind farm developments in Europe were lagging compared to that of the United States. High population density and geographical conditions presented obvious impediments to the proliferation of land-based wind power in Europe, and the first commercial onshore wind turbine was not installed in Europe until 1991.

Crucially, however, it had been well understood for years by European scientists and engineers that wind turbines placed offshore could produce more energy than those located onshore due to stronger, unobstructed wind resources and a lack of man-made and geographical obstacles for the creation of high wind speeds (Nikolaos 2004). A strong motivation for offshore wind in Europe was the considerably higher, steadier, and unobstructed winds met in the open sea, which typically exceed 8 m/s (Kaldellis and Kapsali 2013), tangibly higher than the 3–5 m/s resource range typically found over land. Moreover, wind farms installed offshore would avert the land-use and visual impact issues that often obstructed the development of onshore wind farms.

However, the initial development of offshore wind projects in Europe was driven primarily by the commercial response to the scarcity of suitable onshore sites and conflicts in land-use issues (Kaldellis and Kapsali 2013). Even prior to the installation of the first commercial land-based wind turbine in Europe in 1991, the first commercial wind turbine in Europe was installed offshore. The Nogersund-Savante project, located some 250 metres off the southeastern coast of Sweden in the Baltic Sea, consisted of one 25 kW turbine generator with a rotor diameter of 25 metres. This was followed shortly thereafter in 1991 with the construction of the world's first offshore wind farm array, Vindeby, in Denmark. Constructed in shallow waters (2–6 metres depth) at a distance of some 1.5–3 km offshore the Danish coastal island of Lolland (near the village of Vindeby), the array counted eleven 450 kW wind turbines and a nameplate capacity of 4.95 MW.

Vindeby was followed by a series of other small offshore wind projects throughout Denmark, the United Kingdom, and Holland during the remainder of the decade, each installed at less than 7 km from shore and in water depths of under 8 metres. In 2000, the offshore construction began on what is considered the first large-scale offshore wind farm, Middelgrunden, positioned 2 km off the harbour of Copenhagen. Although this 40 MW project (20 turbines at 2 MW each) helped pave the way for two other large offshore wind projects (the 160 MW Horns Rev 1 project in 2002 and the 165.2 MW Nysted project in 2003), the primary legacy of these handful of pioneering projects was construction cost overruns and frequent turbine equipment failures in the harsh marine environment offshore (Kaldellis and Kapsali 2013). This reputation stifled political and social acceptance of the technology 'for a respectable period of time' (Kaldellis and Kapsali 2013).

1.3   Role of Offshore Wind

The higher construction costs offshore and persistent equipment failures in the marine environment made offshore wind farms more expensive and less reliable than their onshore counterparts. Although initial commercial projects were sporadically popping up in northern Europe, the cost overruns and equipment failures did little to win support for the industry. In the United Kingdom, the prevailing viewpoint throughout the 1990s was that offshore wind was prohibitively expensive and that the technologies wouldn't be economically viable until after 2020. As such, political and commercial support for the industry was generally equivocating, and most government R&D funding support instead concentrated on the technologies deemed closer to being commercially viable (Kern et al. 2014).

Given this context, it was imperative that offshore wind could construct a narrative that would fulfil broader political and social goals in order to garner stronger political and social support across Europe. Fortunately, within the first decade of the 2000s, a series of practical and political considerations would intersect to create a more appealing rationale for the government support of offshore wind in Europe. Notably, in 2002, the EU ratified the Kyoto Protocol, which would eventually impose binding national renewable energy targets onto each EU member state through 2020. Moreover, gas contract disputes between Russia's Gazprom and Ukraine in 2006 and again in 2008–2009 heightened Europe's energy security concerns, particularly with respect to the block's dependency on imported natural gas. Then, the financial crisis of 2008 and subsequent deep recession across the Eurozone lead to renewed political emphasis on jobs creation and reviving non-service industries and, in particular, heavy industry.

As such, a confluence of policy goals stemming from Kyoto Protocol and subsequent Renewable Energy Targets, and practical considerations such as land scarcity and energy security of supply issues, convened to garner palpable political and public support for offshore wind energy in the first decade of the 2000s throughout northern and western Europe. These policy divers are summarised in Table 1.1.

Greenhouse Gas Emissions Reduction Targets GHG mitigation and the EU's commitment to the Kyoto Protocol (including the pursuant 2009 Energy Package) proved to be the most influential policy drivers for the promotion of the offshore wind industry. These political mandates would result in specific capacity targets and policy mechanisms for offshore wind that would effectively pave the way for the development of the industry.

The Kyoto Protocol was ratified by the EU in 2002 and went into effect in 2005. The original framework obliged signatories to reduce

Table 1.1   Political and practical drivers for offshore wind

Major policy drivers	Greenhouse gas emissions reduction and meeting renewable energy targets, Bolstering energy security, Supporting industry and jobs creation	Practical considerations	Land scarcity, Public concerns (visual impact, land-use issues), Greater resource potential offshore

GHG emissions an average of 5% against 1990 levels during the first commitment period, lasting from 2008 through 2012. The EU was not only willing to commit to the Kyoto Protocol, but also eager to implement even more ambitious targets. The EU obliged most of its member states to a GHG reduction of 8% on 1990 levels during the first commitment period[2] and announced as early as 2007 that it would set a series of more aggressive climate targets for 2020. The *2020 Climate Energy Package*, as it became known, was enacted into legislation in 2009 and set forth EU targets of:

- A 20% reduction in GHG emissions compared to 1990 levels
- About 20% of final energy consumption to be provided by renewable energy sources
- An increase in energy efficiency of 20%

The most crucial feature of the Energy Package legislation for the development of an offshore wind industry was that targets on the EU level were handed down into binding national-level targets for each EU member state. The 2020 renewable energy targets were based on available natural resources and other considerations distinctive to each energy market and ranged from as low as 10% (Malta) to as high as 49% (Sweden).

Pursuant to, and in accordance with, this mandate, EU member states have since been required to submit National Renewable Energy Action Plans (NREAPs) to the European Commission that detail the strategy and pathway the country will take to achieve its 2020 targets under the Renewable Energy Directive. The NREAPs must include individual renewable energy targets by sector and detail the planned contribution that different renewable energy technologies are envisaged to make towards fulfilling the overall renewable energy target.

Given the vast offshore wind energy resources available to many major European countries, and the scalability of the technology to produce utility-scale power output, NREAPs for various European countries have established strong goals for offshore wind power. In 2015, the total contribution that installed offshore wind energy capacity was envisaged to make towards countries' renewable energy targets totalled just under 15,000 MW. By 2020, the anticipated contribution of offshore wind energy across Europe increases to an indicative cumulative total of over 41,000 MW. Table 1.2 presents the contribution of offshore wind in Europe NREAPs.

Table 1.2 Contribution of offshore wind in Europe NREAPs

Country	2015 target (MW)	2020 target (MW)
United Kingdom	5500	12,990
Germany	3000	10,000
France	2667	6000
Belgium	1285	2000
The Netherlands	1178	5178
Denmark	661	1339
Ireland	252	555
Spain	150	3000
Sweden	129	182
Portugal	25	75

Source: European Commission (2017) and EWEA (2015)

1.4   Role of Offshore Wind Outside of Europe

Outside of the EU, a growing number of countries have also explored the prospect of offshore wind to help reduce GHG emissions, improve air quality (particularly in major urban areas), and bolster electricity production from carbon-free renewable energy sources.

The distribution of populations globally and the rapid urbanisation to the large coastal cities around the world offer a keen rationale for the role of offshore wind power in many developed and developing economies. This is due to the demographic that large population and industrial centres tend to be concentrated around coastlines and major port cities around the world. In the EU, over 40% of the population resides in coastal regions, which are also key areas for industry. Over half of the United States' power demand comes from coastal areas. In China, over half of the country's 1.3 billion people already live along the coastline, and rapid urbanisation is expected to accelerate this trend over the coming years.

Given these dynamics, it follows that energy demand in these regions is higher and the stress on local power infrastructure is greater. Land-based renewable energy generation sources (such as land-based wind) are inherently limited by availability of land in these densely populated regions and confront bottlenecks in terrestrial transmission capacity and land-use conflicts. As population in these regions increases, these constraints will become more pronounced. As such, for many of the same reasons that Europe first started to experiment with offshore wind farms over 25 years ago, both developed and emerging countries around the world, particularly

Table 1.3 Global offshore wind capacity targets by country

Country	Official offshore wind target
Japan	30 GW by 2050
China	10 GW by 2020
Taiwan	600 MW by 2020/4 GW by 2030
United States	No federal official target set. Some state targets
South Korea	1.5 GW by 2020
India	1 GW by 2020

Source: Authors compiled from local sources

with dense coastal populations and limited land availability, are increasingly interested in establishing offshore wind power generation plants. A summary of offshore wind targets globally is presented in Table 1.3.

Trends in the ever-changing global energy landscape will also continue to shape both the political and public acceptance for the support of new renewable technologies and offshore wind power. Across the developed economies, scores of aging fossil fuel plants and nuclear plants are slated to be decommissioned over the next decade, whilst at the same time investment in carbon-intensive resources by energy companies are facing mounting opposition from shareholders and local communities due to air pollution concerns. On top of this, following the Fukushima disaster of 2011, public acceptance of nuclear power plants has been diminishing in Japan and other major developed countries, further constraining the options to replace aging power generation plants. There is little doubt that the major Organisation for Economic Co-operation and Development economies, in particular, are in the midst of a formative energy transition.

Developing economies similarly face a shifting tide in power generation technologies. Electricity demand is expected to continue to grow over the next decade in developing and emerging economies; however, the international financial institutions and multilaterals such as the World Bank are reluctant to support new investments in coal power plants. Private investments in carbon-intensive resources can also face local backlash. As such, governments in the emerging and developing economies are receptive to establishing new technologies that are accommodated by the support of international financial institutions in order to develop clean energy infrastructure to meet their future power demand. The Paris Agreement furthermore intends to boost these opportunities.

Table 1.4 Summary table—global offshore wind targets and key policy drivers

Country	Target	Key policy drivers	Policy priority
United Kingdom	20 GW by 2020	Energy security:	♦♦♦
		Decarbonisation:	♦♦♦
		Industrial benefit:	♦♦
		Land scarcity	♦
Germany	6.5 GW by 2020	Energy security:	♦♦♦
		Decarbonisation:	♦♦♦
		Industrial benefit:	♦♦♦
		Land scarcity	
Denmark	2.8 GW by 2020	Energy security:	♦♦
		Decarbonisation:	♦♦♦
		Industrial benefit:	♦♦♦
		Land scarcity	♦♦♦
The Netherlands	4.5 GW by 2023	Energy security:	♦
		Decarbonisation:	♦♦♦
		Industrial benefit:	♦♦♦
Belgium	1.8 GW by 2020	Energy security:	♦♦
		Decarbonisation:	♦♦♦
		Industrial benefit:	♦♦
		Land scarcity	♦♦
France	3 GW by 2023	Energy security:	♦
		Decarbonisation:	♦♦♦
		Industrial benefit:	♦♦♦
		Land scarcity	
PR of China	10 GW by 2020	Energy security:	♦♦♦
		Decarbonisation:	♦♦♦
		Industrial benefit:	♦♦
		Land scarcity	♦♦
		Air quality	♦♦♦
Japan	30 GW by 2050	Energy security:	♦♦♦
		Decarbonisation:	♦♦
		Industrial benefit:	♦
		Land scarcity	♦♦♦
Taiwan	600 MW by 2020, 4 GW by 2030	Energy security:	♦♦♦
		Decarbonisation:	♦
		Industrial benefit:	♦♦
		Land scarcity	♦♦
India	1 GW by 2020	Energy security:	♦♦♦
		Decarbonisation:	♦♦♦
		Industrial benefit:	♦♦♦
		Land scarcity	♦♦
		Air quality	♦♦

(*continued*)

Table 1.4   (continued)

Country	Target	Key policy drivers	Policy priority
United States	No official federal target	Energy security: Decarbonisation: Industrial benefit: Land scarcity	◆ ◆

Source: Authors adapted from Shukla et al. (2015)

Within this context, the role that offshore wind energy can play in a country's political, economic, and social landscape is gaining increasingly widespread attention, both in developed and developing countries. Table 1.4 presents global offshore wind targets and key policy targets.

1.5   CONCLUSIONS

The development of offshore wind energy has benefitted from the growing political and public acceptance that accompanies renewable energy, particularly as governments implement policies to mitigate climate change by curbing GHG emissions. However, offshore wind has also carved out a unique role in the political context and public opinion of a growing number of countries as of late. In addition to bolstering energy security and decarbonising the power sector, offshore wind farms make use of a sea of otherwise untapped wind energy resources located just offshore many countries' largest and fastest growing demand centres. As the stresses from urbanisation and growing electricity demand in many parts of the world continue to mount, offshore wind technology has the potential to meet these challenges.

Nevertheless, developing offshore wind energy remains costly and technically complex, and must compete with and continuously be justified in the context of several other low- and zero-carbon energy alternatives. The following chapters detail the current global market for offshore wind and seek to dissect the technical challenges and cost drivers, public acceptance issues, and evaluate the policy frameworks and support mechanisms that are currently being applied to offshore wind markets globally, and discuss the lessons learned thus far, and implications for policy makers in pursuing offshore wind.

NOTES

1. It should also be noted that the same year, 1887, in Cleveland, Ohio (USA), a wind-driven turbine generator produced by inventor and industrialist Charles Bush was also erected and supplied power to his home. However, Blyth's turbine at Marykirk is widely considered to be the first wind-driven electricity generator.
2. The states which were members of the EU before 2004 must collectively reduce their greenhouse gas emissions by 8% between 2008 and 2012 with the exception of Poland and Hungary (6%), and Malta and Cyprus.

REFERENCES

European Commission. (2017). *National Action Plans.* Available online at https://ec.europa.eu/energy/en/topics/renewable-energy/national-action-plans. Accessed 5 Oct 2017.

EWEA. (2015). *The European Offshore Wind Industry—Key Trends and Statistics 2015.* European Wind Energy Association, February 2016. https://www.ewea.org/fileadmin/files/library/publications/statistics/EWEA-European-Offshore-Statistics-2015.pdf

Kaldellis, J. K., & Kapsali, M. (2013). Shifting Towards Offshore Wind Energy—Recent Activity and Future Development. *Energy Policy, 53,* 136–148.

Kern, F., Smith, A., Shaw, C., Raven, R., & Verhees, B. (2014). From Laggard to Leader: Explaining Offshore Wind Developments in the UK. *Energy Policy, 69,* 635–646.

Nikolaos, N. (2004). *Deep Water Offshore Wind Technologies.* MA thesis University of Strathclyde. Retrieved from. http://www.esru.strath.ac.uk/Documents/MSc_2004/nikolaos.pdf

Rivkin, D. A., Randall, M., & Silk, L. (2014). *Wind Power Generation and Distribution.* USA: Jones and Bartlett Learning.

Shahan, Z. (2014, November 21). *History of Wind Turbines. Renewable Energy World.* http://www.renewableenergyworld.com/ugc/articles/2014/11/history-of-wind-turbines.html. Accessed 28 Apr 2017.

Shukla, S., Reynolds, P., & Felicity Jones, F. (2015). *Offshore Wind Policy Drivers in Europe and China.* Indian Wind Power. Global Wind Energy Council (GWEC). Retrieved from http://www.gwec.net/wp-content/uploads/2015/05/Indian_Wind_Power-April-May_2015.pdf

Stenkjaer, N. (2009, July). *Poul la Cour—The Danish Wind Turbine Pioneer.* Nordic Folkecenter. Retrieved from http://www.folkecenter.net/gb/rd/wind-energy/48007/poullacour/

Swift-Hook, D. T. (2012). History of Wind Power. In A. Sayigh (Ed.), *Comprehensive Renewable Energy* (1st ed., Vol. 2, pp. 41–72). Amsterdam: Elsevier, ISBN: 9780080878737.

CHAPTER 2

Global Offshore Wind Market

Abstract The global market for offshore wind power has rapidly evolved over the past decade. Once considered to be a purely European development, the offshore wind industry is now expanding into virtually all corners of the globe. Far East Asia, North and South America, the Indian subcontinent, and Oceania, among others, are all making strides towards developing this resource in their waters. This chapter discusses the latest development in the global market. It also explores the potential for growth in not only the aforementioned geographic areas, but also other, yet unexplored, opportunities.

Keywords Offshore wind market structure • Offshore wind global capacity • Offshore wind expansion • European offshore wind market • North American offshore wind market • East Asian offshore wind market

2.1 INTRODUCTION

Although global power generation is dominated by fossil fuels such as coal, oil, and natural gas, electricity markets around the globe are in the process of rapid transformation. The share of renewables is swiftly increasing in the generation mix and this trend is expected to continue in the future. According to Bloomberg New Energy Finance (2017), between

© The Author(s) 2017
R. Poudineh et al., *Economics of Offshore Wind Power*,
https://doi.org/10.1007/978-3-319-66420-0_2

now and 2040, around \$7.4 trillion, or three quarters of new investments in power generation worldwide, will be spent on wind and solar technologies. This provides a tremendous opportunity for the offshore wind industry to absorb some of this capital and secure its position in the global generation mix.

Currently, the offshore wind industry is expanding to all areas of the globe. Once considered a wholly European enterprise, this is decidedly no longer the case. The supply chain in Europe has been branching out to nascent markets and lending their expertise in all matters technical and political, including the very real consideration of all stakeholders invested in our planet's ocean resources. China has now become the owner of the third largest offshore wind fleet when measured in terms of nameplate capacity. The United States and Korea have recently started producing power from their first commercial-scale wind farms. Australia, a historically coal-driven energy economy, is now starting to explore offshore wind as well.

Although the global offshore wind market is rooted in Europe and global trends are largely based on the European experience, a wider distribution of development would have several advantages to the industry. For instance, the future Asian supply chain could provide a counterweight to the European suppliers and increase the competition within the industry. Furthermore, it can create significant new experiences, knowledge, and opportunities for the development of new technologies in relation to foundation and turbine design, as site conditions in many Asian markets are different from those in Europe.

Along with global expansion, the offshore wind industry can also benefit from its shared activities with the offshore oil and gas sector which is a mature industry. This includes, for example, installation and operation of assets in the risky marine environment. In recent years, major oil and gas companies have been entering this business in order to leverage their capital, expertise, and technical knowledge in the area of offshore operation. However, the extent of benefits for offshore wind from the shared features of these two supply chains is dependent on myriad factors. Foremost among these is the amount of activity in the offshore oil and gas sector which is directly correlated with the global price of oil.

This chapter explores the current installed global capacity by market in the ensuing section. Section 2.3 then delineates how offshore wind fits into the global power generation mix, and then, lastly, the points are summarised in Sect. 2.4.

2.2   INSTALLED GLOBAL CAPACITY BY MARKET

As of year-end 2016, there was some 14,384 MW of offshore wind capacity installed globally (GWEC 2017). European markets currently account for nearly 88% of globally installed offshore wind capacity, or 12,631 MW. By the end of 2016, there were 3589 offshore wind turbines installed and grid-connected across 81 wind farms in European waters spread across ten countries (WindEurope 2017).

It is estimated that offshore wind generation capacity in Europe will effectively double to 24,600 MW by 2020 as projects currently under construction and in planning become grid-connected (WindEurope 2017). The United Kingdom is the predominant market leader in offshore wind, with approximately 5100 MW of grid-connected capacity and a target of reaching some 10,000 MW by 2020. Germany has been narrowing this gap significantly over 2015 and 2016 and now has over 4000 MW of installed capacity. China and Denmark follow in terms of capacity (see Fig. 2.1), each boasting over 1 GW of installed capacity. Denmark, Germany, and the United Kingdom together represent over 80% of Europe's total installed capacity.

However, a catalogue of obstacles (such as fluctuating incentive and support models in key markets such as the United Kingdom and Germany and delayed grid expansion or connection plans) has hindered the progress of offshore wind developments at different points over the past half-decade. An uneven rollout of policies or bottlenecks in grid connectivity has led major markets such as the United Kingdom and Germany to falter for some substantial periods in their development. As a result, the indicative EU 2020 target of over 40 GW installed (based on the EU member states' National Renewable Energy Action Plans [NREAPs]) is very unlikely to be reached. Moreover, although the target of 24,600 MW, as stated by WindEurope, is a significant reduction from the cumulative EU target of 40 GW, even this revised trajectory entails doubling the currently installed capacity in the four years between 2017 and 2020. Nevertheless, despite being cumulatively behind targets, most European countries have shown steadily increasing offshore wind capacity, and the installation of offshore wind farms in key European countries is also slated to accelerate in the coming years.

The development of Europe's offshore wind markets over the past several decades has rendered Europe the de facto authority in the offshore wind industry. Denmark, Germany, and Spain are the global hubs for off-

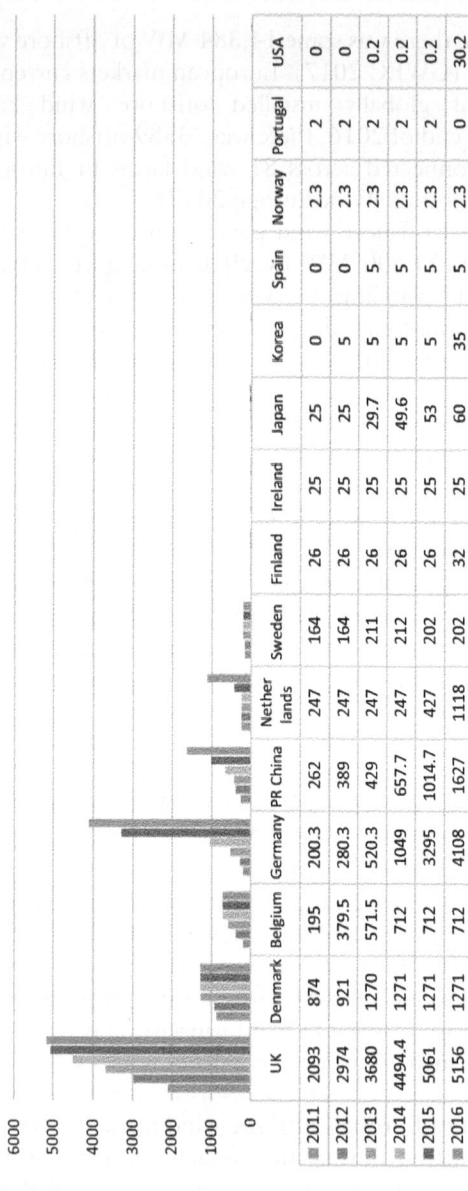

	UK	Denmark	Belgium	Germany	PR China	Netherlands	Sweden	Finland	Ireland	Japan	Korea	Spain	Norway	Portugal	USA
2011	2093	874	195	200.3	262	247	164	26	25	25	0	0	2.3	2	0
2012	2974	921	379.5	280.3	389	247	164	26	25	25	5	0	2.3	2	0
2013	3680	1270	571.5	520.3	429	247	211	26	25	29.7	5	5	2.3	2	0.2
2014	4494.4	1271	712	1049	657.7	247	212	26	25	49.6	5	5	2.3	2	0.2
2015	5061	1271	712	3295	1014.7	427	202	26	25	53	35	5	2.3	2	0.2
2016	5156	1271	712	4108	1627	1118	202	32	25	60	35	5	2.3	0	30.2

Fig. 2.1 Installed offshore wind capacity from 2011 to 2016 (MW) (Source: Authors compiled from (EWEA 2015); (WindEurope, 2016); (GWEC 2014); (GWEC 2015); (GWEC 2016))

shore wind turbine technology and manufacturing. European countries house the leading offshore wind companies, purpose-built installation vessels, technology providers, fabrication and installation experience, and know-how. Outside of Europe, installed offshore wind capacity totalled just over 1750 MW by the end of 2016. China accounts for the bulk of this capacity with 1627 MW grid-connected by the end of 2016. China is followed distantly by Japan (60 MW), South Korea (35 MW), and the United States (30 MW) (GWEC 2017). Vietnam has developed a 99.2 MW project in inter-tidal waters and is working towards developing true offshore projects. Taiwan, India, and Canada have all made progress in the offshore wind industry; however, they are yet to have utility-scale offshore wind farms installed.

2.2.1   Infrastructure and Supply Chain Development

The existing oil and gas infrastructure, in close proximity to strong wind resources in northwest Europe, has been a significant boon to the offshore wind industry. The decline of oil and gas production in the North Sea, coupled with declining oil prices in the mid-2010s, has also proven advantageous to the offshore wind industrial supply chain in northwest Europe. Traditional servicers of the oil and gas industry are increasingly retrofitting facilities to accommodate offshore wind in order to diversify from oil and gas, thus securing the foundation for stronger development potential and competition in the offshore wind industry. Many of the original installation vessels for oil and gas have been borrowed by the offshore wind sector to install massive wind turbines and substructures offshore. Heavy industrial fabrication yards, typically dedicated to oil and gas, are now ramping up investments in facilities dedicated to the streamlined serial manufacture of offshore wind structures. These developments have proven both timely and necessary for the growth of Europe's offshore wind industry.

However, the migration of resources from oil and gas to the offshore wind industry is, in many ways, unique to the European landscape. The commercial case for diverting capital and resources away from the oil and gas sector, or from the even more economical land-based renewables such as onshore wind power, is far less logical in many markets outside of Western Europe. In the United States, for example, the oil and gas infrastructure remains concentrated around the US Gulf of Mexico, disconnected from the strong coastal wind resources and offshore wind lease areas found along the US east and west coasts. Despite the protracted slump in the oil and gas industry, Gulf Coast asset owners nevertheless remain reluctant to mobilise resources to offshore wind areas until such a time as there is a proven pipeline of projects to justify such a large capital investment. Otherwise, there is simply no strong commercial rationale for the support of such an investment from the private sector. Outside of Europe, many countries are thus locked into what has become akin to a 'chicken and egg' dilemma. Project developers insist on a local supply chain to commit to building a pipeline of projects; however, the local supply chain insists on a pipeline of projects before making capital investments in the tailored supply chain. As discussed in later chapters, this presents substantial opportunity for policy makers to address.

2.2.2   Project Development and Investment

Commercial-scale offshore wind projects have traditionally been developed in Europe primarily by large utilities and power producers such as Vattenfall, RWE, E.ON, Centrica, and SSE as well as a few major oil and gas producers such as Danish Oil and Natural Gas (DONG). More recently, a number of special-purpose joint venture entities have also been formed by utilities to develop projects. By 2016, 67% of the equity investors in the industry were represented by major power producers (Fig. 2.2).

These large players have typically been able to leverage assets on their balance sheets to self-finance projects. However, the appetite and capability of equity investors, including major utilities, to finance off the corporate balance sheet have been in retreat given the escalation of project costs. To some extent, these escalations have materialised because the scale of the offshore wind farms is getting substantially larger as is necessary to economise projects. In Europe, the average size of an offshore wind farm under construction jumped 12.5% from 2015 to 2016 to reach some 379.5 MW (WindEurope 2017). The sites are moreover moving further from shore, which also drives up the cost of offshore construction. The escalation of these costs has at least two major implications. First, it means that the ability to meet aggressive national targets for offshore wind deployment in Europe will require hefty capital investments. It was recently figured that meeting a cumulative EU target of some 40 GW of offshore wind capacity would require at least €110 billion in additional capital by the end of the decade (Boston Consulting Group 2013). Second, these trends suggest that if the projects cannot be self-financed by

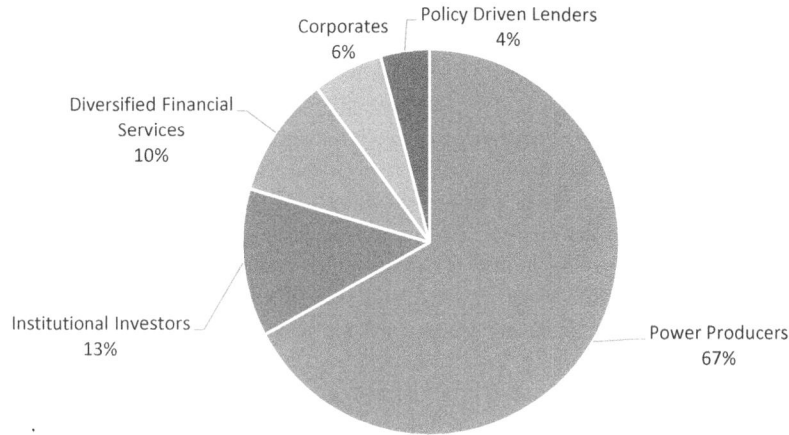

Fig. 2.2 Major equity investors (2016) (Source: WindEurope (2017))

utilities or corporate financing (as discussed in greater details in Chap. 6), then new sources of capital from a wider fleet of investment vehicles, both private and public, will be necessary to keep the industry viable.

The profile of equity investors in offshore wind developments in Europe has thus been shifting. In 2016, project owners in Europe sold or divested some 2800 MW of offshore wind power projects. These numbers are similar to those of 2015, but both of these years were a significant increase from the years preceding. Four out of the top five biggest acquirers of these assets have been international investors. The financial services industry, including institutional funds, are also starting to increasingly take equity stakes in Europe's offshore wind farms (WindEurope 2017). As discussed in later chapters, the changing ownership and investor profile of European offshore wind farms could have a bearing on project financing, the flow of capital into the industry, and the cost of capital. These areas can ultimately impact the competitiveness of the offshore wind industry.

2.3   Offshore Wind in the Global Energy Mix

2.3.1   Renewable Energy in Global Power Generation

In order to become a truly global mainstream generation source, offshore wind technology must prove itself to be cost competitive with other sources of power generation, including both fossil fuels and other renewable generation technologies. Global power generation continues to be dominated by coal, natural gas, and oil-based generation sources (diesel, fuel oils, etc.). In 2015, the demand for coal, oil, and natural gas accounted for over 85% of total global primary energy demand (Fig. 2.3). To date, renewable generation sources still only satisfy a fraction of global primary energy consumption. Excluding hydroelectricity, power generation from renewable energy sources (including wind, solar, biomass, geothermal, and waste) accounted for just 3% of global primary energy demand and just under 7% of global power generation as of 2015 (BP 2016). Figure 2.4 shows the share of renewables, based on technology type, from the global electricity production.

Nevertheless, power markets are in a process of a rapid structural transition. This is foremost because policy makers are seeking to incentivise the uptake of renewables, particularly in the power generation sector, and the cost of renewable technologies is also becoming more competitive with certain types of fossil fuel power plants. As evidenced, the use of oil products (such as diesel) for power generation is in long-term structural

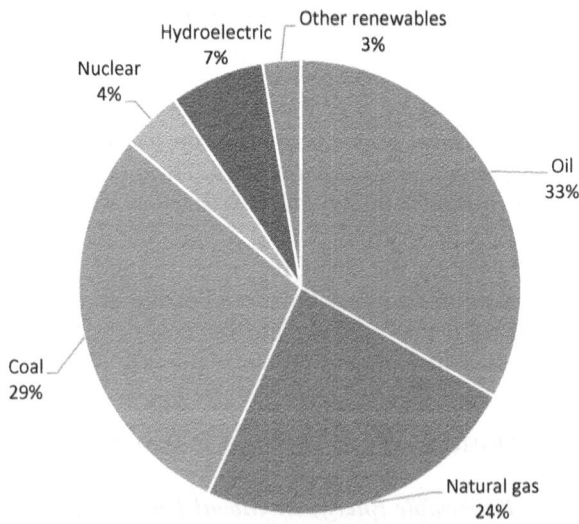

Fig. 2.3 Total primary energy consumption 2015 (Source of data: BP (2016))

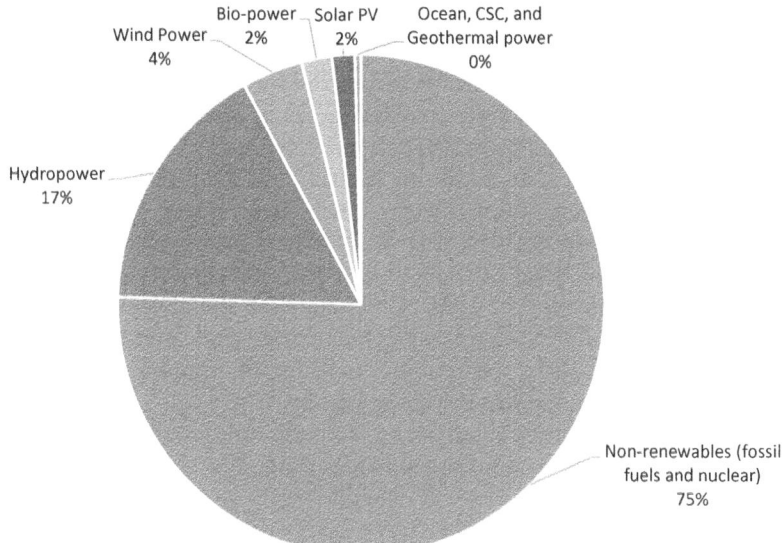

Fig. 2.4 Global electricity production 2016 (Source of data: REN21 (2017))

decline as they are typically too expensive relative to alternatives and, moreover, expose countries to energy security and commodity price fluctuation risks. Coal, on the other hand, is typically cheap and global demand remains strong. However, the commodity faces increasing regulatory pressure in a number of markets that will likely add to its generation costs in the long run and erode its economic advantage over efficient renewables. Several countries in Europe and elsewhere have also recently committed to phasing out coal for electricity generation, and, as previously stated, many multilateral and other funding institutions are no longer financing coal use projects (REN21 2017).

Governments are also becoming increasingly adept at integrating higher shares of variable renewable energy into their base load power supply. As described in subsequent chapters, this is being achieved largely through improved resource forecasting, electricity storage, demand response, and coordination and trade of electricity supply across larger balancing areas (REN21 2017). As a result, variable renewable energies are routinely able to meet over 25% electricity demand in several European countries.

The year 2013 also marked the first year when more renewable generation capacity was added globally than for fossil fuel plants (e.g., coal, natural gas, and oil) combined. This trend has continued unabated since then.
The most recent data shows that 161 GW of renewable energy generation
capacity was brought online in 2016, which accounted for 62% of all net
power generation capacity additions globally (REN21 2017). In total,
renewable energy generation capacity (including hydropower) reached
2017 GW by the end of 2016 (Fig. 2.5). When excluding hydropower,
the total capacity of installed renewable generation capacity totalled just
over 920 GW by the end of 2016. Chief amongst the growth of renewables has been utility-scale solar PV technology and land-based wind
energy. At present, these technologies account for roughly 85% of the
globally installed renewable generation capacity.

In the midst of what, by all accounts, is a renewable energy boom, however, the role of offshore wind remains uncertain. The impact of offshore
wind in the global energy mix will depend largely on how effectively the
industry can reduce its cost curve to be competitive not only with fossil
fuels, but other renewable energy technologies. Often-referenced
measurements of plants' economics currently suggest that offshore wind

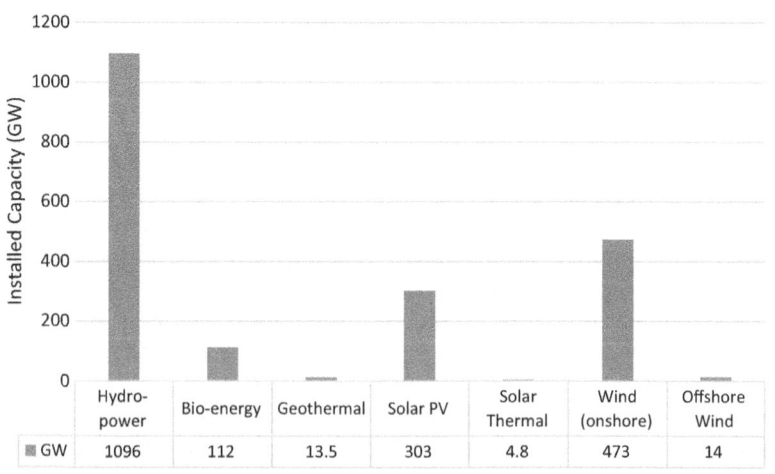

Fig. 2.5 Global total renewable generation capacity by technology (2016)
(Source: Authors derived from REN21 (2017))

plants, on average, are far from being cost competitive with both fossil fuel generation sources (with the exception of diesel), as well as other conventional renewable energy technologies. As a result, other generation technologies such as solar PV and onshore wind have experienced steady flows of private capital and huge capacity gains in recent years, whereas (globally) the offshore wind industry has suffered ebbs and flows in private investment due to high relative development costs and uncertain or insufficient policy support frameworks. This will have a bearing on the development potential for offshore wind. Indeed, since the installation of the first offshore wind farm in Denmark in 1991 (Vindeby), the development of global offshore wind capacity has been just 2% that of land-based wind (ShrutiShukla 2014). Even in the EU, offshore wind capacity is just 9% that of land-based wind (GWEC 2017). As a new technology and industry, the lifetime plant cost of an offshore wind farm exceeds those for nearly every other generation source, and this largely explains its lacklustre development on the global scale when compared to more mature renewables. An analysis of the typical cost comparison methodology is provided in detail in the following section.

2.3.2   *Levelised Cost of Energy and Plant Costs Comparison*

The levelised cost of energy (LCOE) is often used to evaluate and compare the costs of electricity generation for a given plant. The formula is able to take into account plant-level effects from technology design changes, fixed costs, and other inputs. Although methodologies vary, the calculation typically incorporates four major inputs of the plant: installed capital cost (CAPEX), annual operating cost (OPEX), annual energy production, and the fixed charge rate (a coefficient that expresses the cost of financing over the plant's operational life). An example of an LCOE equation is found below (Krupa and Poudineh 2017):

$$
\mathrm{LCOE} = \frac{\sum_{t=1}^{n} \dfrac{I_{t} + O\,\&\,M_{t} + F_{t}}{(1+r)^{t}}}{\sum_{t=1}^{n} \dfrac{E_{t}}{(1+r)^{t}}}
$$

where:

I_t = Investment in year t ($\$/kW/year$)
$O\&M_t$ = Operations and maintenance (O&M) ($\$/kW/year$)
F_t = Fuel cost ($\$/kW/year$)
E = Electricity output (kWh/kW/year)
r = discount rate
t = lifespan (years of the project)

As stated, the LCOE figure is widely referenced when comparing plant-level economics and technologies against one another. Nevertheless, some of the limitations of this approach should be noted. The LCOE estimation ignores the cost of integration (network reinforcement, backup generation, and storage requirements) relating to a particular technology. Such costs are likely to become more important as penetration of offshore wind power increases. In this case, LCOE can be a misleading metric for comparing the attractiveness of offshore wind with other technologies. Furthermore, LCOE is only the measure of cost and does not say anything about profitability and competitiveness, which are related to 'market value' rather than LCOE. As demand for electricity varies continuously and storage is costly, the value of electricity—reflected in price—fluctuates continuously depending on the demand and supply condition. For example, if offshore wind is generating power when and where it has highest value, then a plant's economics may be better than that suggested by its LCOE value. Conversely, if generation from a wind source occurs when it has a low market value and where it imposes high transmission costs, it may be less attractive than that plant's LCOE might suggest. In some markets, periods of high wind generation coincides with very low spot market prices (Cutler et al. 2011).

This notwithstanding, for regulatory and policy purposes, the LCOE figure for offshore wind is often referenced—not only to calculate subsidies and feed-in tariff levels, but also as a basis of comparison against other power generation plants—despite its inherent limitations. Figure 2.6 demonstrates the wide disparities between offshore wind and other electricity generation sources on an actual cost basis.

It should be noted that the LCOE for land-based wind has retreated dramatically in recent years, contracting some 58% over the five years between 2009 and 2014 (Lazard 2014). In some markets, the LCOE of land-based wind now averages as low as $\$32/MWh$, cheaper than many fossil fuel plants.

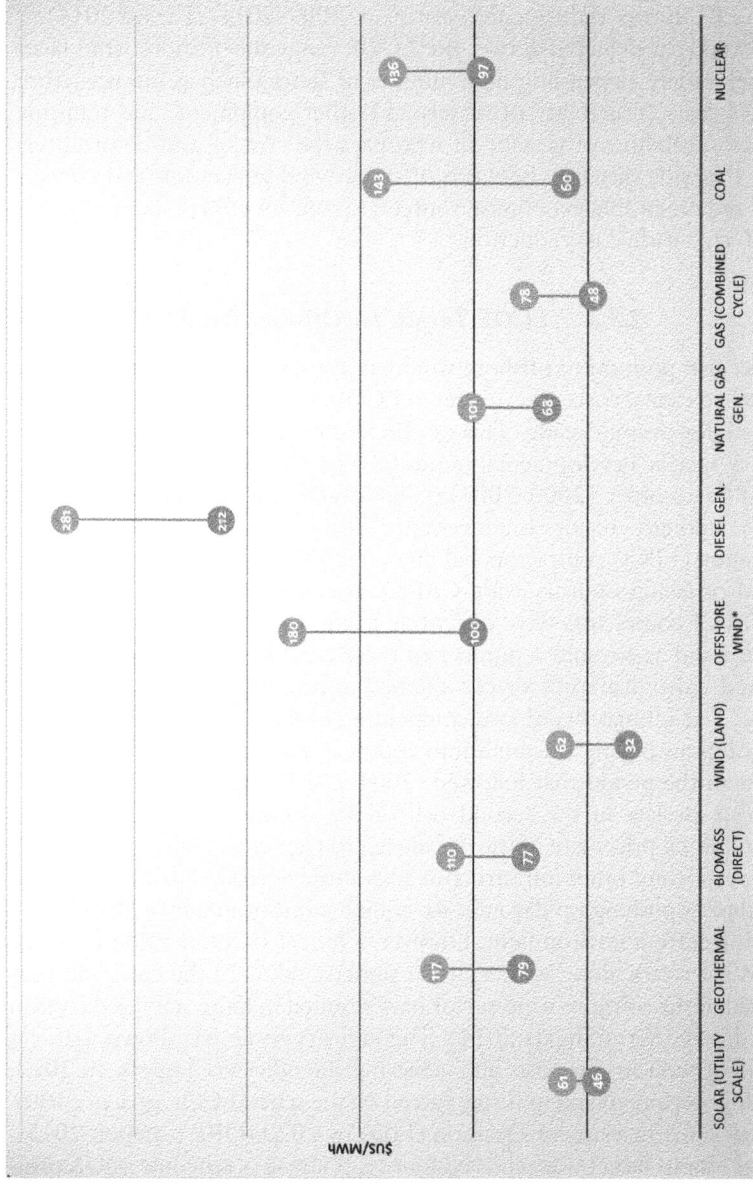

Fig. 2.6 LCOE comparison across different technologies—$/MWh (in US$) (Source: Authors compiled from Lazard (2016). *Authors' estimate based on various sources and reports)

Efficiencies in solar PV have been even more pronounced, with a 78% reduction in LCOE over the same five-year span (2009–2014) (Lazard 2014).

However, as depicted above, the LCOE range for offshore wind farms can vary widely, depending on a number of factors such as infrastructure, logistics costs, availability of vessels and other equipment, and transmission responsibilities (the issue of who pays the cost of grid connection). Given the wide disparity between offshore wind generation and conventional and renewable generation sources, significant effort is being focused on industry-wide cost reduction.

2.3.3   LCOE Targets for Offshore Wind

The peculiar position of offshore wind has indeed been due to the fact that its installed capital costs, a key driver in LCOE, have continued to rise, rather than fall, for over a decade. This can be attributed to a number of unique, industry-specific developmental trends (GWEC 2014). Notably, in the early part of this century (2000–2004), initial overzealousness and 'competitive hustle' amongst major Engineering, Procurement, Construction, and Installation (EPCI) contractors bidding on offshore wind tenders resulted in a trend of falling industry-wide CAPEX averages. However, in hindsight, these EPCI contractors were effectively failing to anticipate costs and risks properly, and as a result, a number of these contractors and suppliers were rendered either bankrupt or out of the business by the mid-point of the decade. This ultimately reduced competition in the European sector, and as more cautious pricing was built into contracts amongst the remaining contractors in the period that followed (2004–2010), average project CAPEX began to escalate in the second half of the decade. Adding to this, the rebound in oil prices late in the decade led to restricted availability of installation vessels and other infrastructure previously borrowed from the oil and gas industry, pushing up day rates for equipment (ShrutiShukla 2014).

In the current environment, however, a retreat in competition from the oil and gas sector along with a clearer understanding of the costs and risks involved in the offshore wind sector have resulted in more stabilised costs in this industry (ShrutiShukla 2014). The industry so far has shown that it is capable of reducing its costs and achieving the efficiency targets. In 2012, the UK government and industry agreed on the stated LCOE goal of £100/ MWh at Final Investment Decision (FID) by 2020 (ORE Catapult 2015). This ambitious target was achieved four years ahead of schedule as UK projects that made an FID in 2015–16 had an average LCOE of £97/MWh, a

32% reduction from £142/MWh for projects reaching FID in 2010–11 (Catapult Offshore Renewable Energy 2017). Furthermore, in the second round of contract for difference (CfD) allocation in 2017, offshore wind in the UK cleared at a price as low as £57.5/MWh. This demonstrates not only that the industry has a high potential for efficiency gain, but that it can cut back costs at a much faster pace than was previously predicted.

2.4   Conclusions

The strong development of Europe's offshore wind markets has rendered Europe the de facto leader in this industry and a global hub for wind turbine technology and manufacturing. The decline of oil and gas production in the North Sea, in proximity to the existing oil and gas infrastructure along with declining oil prices, has allowed the offshore wind industry to borrow many of the original installation vessels for oil and gas in order to install massive wind turbines and substructures. However, once considered a wholly European enterprise, the offshore wind industry is now branching out to nascent markets and lending their expertise in all matters including technical, regulatory, and political issues with offshore wind development. As of 2016, the global offshore wind capacity was distributed among several different geomarkets.

The global expansion of offshore wind energy is part of a larger trend in which renewable energy is gaining share in the total energy supply. Renewable sources have continuously increased as a percentage of added generation each year since 2013, and the trend looks to continue in this direction. As policy initiatives put added pressure on carbon-based energy and technological improvements continue to drastically reduce the cost of renewable power sources, it is likely that the percentages of renewable-based energy will continue to climb to new heights. However, there still exist formidable technical hurdles and challenges to bringing offshore wind costs into alignment with other renewable and carbon-based sources to make it a truly viable, mainstream power generation technology.

References

Bloomberg New Energy Finance. (2017) *New Energy Outlook 2017*. https://about.bnef.com/new-energy-outlook/. Accessed 18 June 2017.

Boston Consulting Group. (2013). *EU 2020 Offshore Wind Targets: The €110 Billion Financing Challenge*. www.bcg.de/documents/file128841.pdf

BP. (2016). *BP Statistical Review of World Energy June 2016*. http://www.bp. com/content/dam/bp/pdf/energy-economics/statistical-review-2016/bp-statistical-review-of-world-energy-2016-full-report.pdf

Catapult Offshore Renewable Energy. (2017). *Cost Reduction Monitoring Framework 2016*. Summary Report to the Offshore Wind Programme Board. Retrieved from https://s3-eu-west-1.amazonaws.com/media.ore.catapult/wp-content/uploads/2017/01/24082709/CRMF-2016-Summary-Report-Print-Version.pdf

Cutler, N. J., Boerema, N. D., MacGill, I. F., & Outhred, H. R. (2011). High Penetration Wind Generation Impacts on Spot Prices in the Australian National Electricity Market. *Energy Policy, 39*, 5939–5949.

EWEA. (2015, January). *The European Offshore Wind Industry—Key Trends and Statistics for 2014*. European Wind Energy Association. www.ewea.org/fileadmin/files/library/publications/statistics/EWEA-European-Offshore-Statistics-2014.pdf. Accessed 21 June.

GWEC. (2014). *Global Wind Annual Market Update. Global Wind Energy Council*. www.gwec.net/wp-content/uploads/2015/03/GWEC_Global_Wind_2014_Report_LR.pdf

GWEC. (2015). *Global Offshore, Global Wind Energy Council*. www.gwec.net/global-figures/global-offshore/. Accessed 20 June.

GWEC. (2016). *Global Wind Report Annual Market Update 2015*. http://www.gwec.net/wp-content/uploads/vip/GWEC-Global-Wind-2015-Report_April-2016_22_04.pdf. Accessed 4 June 2017.

GWEC. (2017). *Global Wind Report Annual Market Update 2016*. http://files.gwec.net/files/GWR2016.pdf. Accessed 4 June 2017.

Krupa, J., & Poudineh, R. (2017). *Financing Renewable Electricity in the Resource-Rich Countries of the Middle East and North Africa: A Review EL 22*. Oxford Institute for Energy Studies. https://www.oxfordenergy.org/publications/financing-renewable-electricity-middle-east-north-africa-review/

Lazard. (2014). *Lazard's Levelized Cost of Energy Analysis—Version 8.0*. www.lazard.com/PDF/Levelized%20Cost%20of%20Energy%20-%20Version%208.0.pdf

Lazard. (2016). *Lazard's Levelized Cost of Energy Analysis—Version 10.0*. https://www.lazard.com/media/438038/levelized-cost-of-energy-v100.pdf

ORE Catapult. (2015, 26 February). *Cost of Offshore Wind Energy Falls Sharply*. https://ore.catapult.org.uk; https://ore.catapult.org.uk/-/cost-of-offshore-wind-energy-falls-sharply-industry-ahead-of-schedule-on-cost-reduction-but-warns-of-challenges-ahead. Accessed 17 June.

REN21. (2017). *Renewables 2017 Global Status Report*. Renewable Energy Policy Network for the 21st Century. http://www.ren21.net/wp-content/uploads/2017/06/170607_GSR_2017_Full_Report.pdf

ShrutiShukla, P. F. (2014, December). *Offshore Wind Policy and Market Assessment: A Global Outlook*. FOWIND, Global Wind Energy Council (GWEC). www.gwec.net/wp-content/uploads/2015/02/FOWIND_offshore_wind_policy_and_market_assessment_15-02-02_LowRes.pdf

WindEurope. (2017, January). *The European Offshore Wind Industry Key Trends and Statistics 2016*. https://windeurope.org/wp-content/uploads/files/about-wind/statistics/WindEurope-Annual-Offshore-Statistics-2016.pdf. Accessed 4 June 2017.

CHAPTER 3

Cost Drivers and Technical Hurdles

Abstract Understanding the specific cost drivers of offshore wind helps to devise effective strategies to lower or remove the barriers to viability and the progress of this industry. This is specifically important given that offshore wind historically has had a higher cost compared to other renewables such as onshore wind. This can be attributed largely to the nature of working in the harsh marine environment, which entails increased risk for projects and assets, and, moreover, demands more performance from those assets as compared to their onshore counterparts. This chapter explores the various cost drivers and technical hurdles of the offshore wind technology.

Keywords Technical hurdles • Cost drivers • Supply chain • European offshore wind market • North American offshore wind market • East Asian offshore wind market

3.1 Introduction

The cost drivers and technical hurdles of the offshore wind industry are major impediments to the smooth penetration of this technology in the global generation mix. The power generated by offshore wind is done so at quite a premium when compared to traditional fossil fuel-based power generation, or even as compared to its renewable counterparts such as

© The Author(s) 2017

R. Poudineh et al., *Economics of Offshore Wind Power*,
https://doi.org/10.1007/978-3-319-66420-0_3

solar photovoltaic and onshore wind (see Chap. 2). The nature of operating in the harsh marine environment mandates that most activities in this sector are restricted by weather conditions. In addition, the sea environment exposes technical equipment to more risk with the ultimate implication of increased cost. This is specifically important in relation to subsea cables which are critical for the output and revenue of wind farms. Adding to this, the issues of disharmony between aerodynamic and hydrodynamic design processes of wind farms and the need for flexibility in the power system to integrate renewables, it shows that the offshore wind industry needs to overcome a significant number of technical challenges before it becomes a mainstream technology.

However, the issues of high costs and technical challenges are not unique to offshore wind energy. Historically, all new technologies and industries are more expensive at the outset. However, once a learning curve is established and supply chains are developed, these costs may trend downward (MIT 2015). Government subsidies are often the linchpin that holds together new industries while they are in their early growth phase. These subsidies are justified on the premise that they can help spur economic growth, and job creation in the long term, incentivise innovation, as well as meet other policy objectives such as decarbonisation or energy security.

An important advantage of understanding the cost drivers and technical hurdles is that it helps to devise effective strategies for lowering the costs and removing the barriers to the progress of the industry. It is also informative about the future costs of development and investments in the sector. Furthermore, it provides an opportunity for the offshore wind industry to borrow from the experiences of mature industries such as offshore oil and gas, which have gone through the same phases of technology development.

This chapter concentrates on cost drivers and technical hurdles which necessarily must be overcome to continue the growth of this industry both in Europe and, indeed, globally. Section 3.2 presents major cost drivers before Sect. 3.3 discusses technical hurdles. Section 3.4 provides the concluding remarks.

3.2   Cost Drivers

The cost divers of offshore wind farms can be categorised by the capital expenditure requirement of planning and constructing the offshore wind farm (CAPEX), as well as the ongoing operational and maintenance costs

(OPEX) associated with the installed wind farm. Each of these categories is broken down into their components below.

3.2.1   *Capital Expenditure (CAPEX)*

CAPEX associated with the offshore wind structure can be broken down into five major project cost centres: the wind turbine generator (WTG) cost, fabrication of the foundations used to anchor the WTG to the seafloor, electrical infrastructure, offshore installation, and planning and development costs (which include permitting and construction financing fees, among other costs). These figures vary significantly market by market depending on existing infrastructure, the availability of installation vessels, and competition amongst industry participants. The European market, as an established industry, is represented by Fig. 3.1. However, these values are constantly in flux as great strides are being made in the European market towards reducing future costs. The emerging markets of East Asia and North America will have unique CAPEX cost breakdowns as they develop their own supply chains.

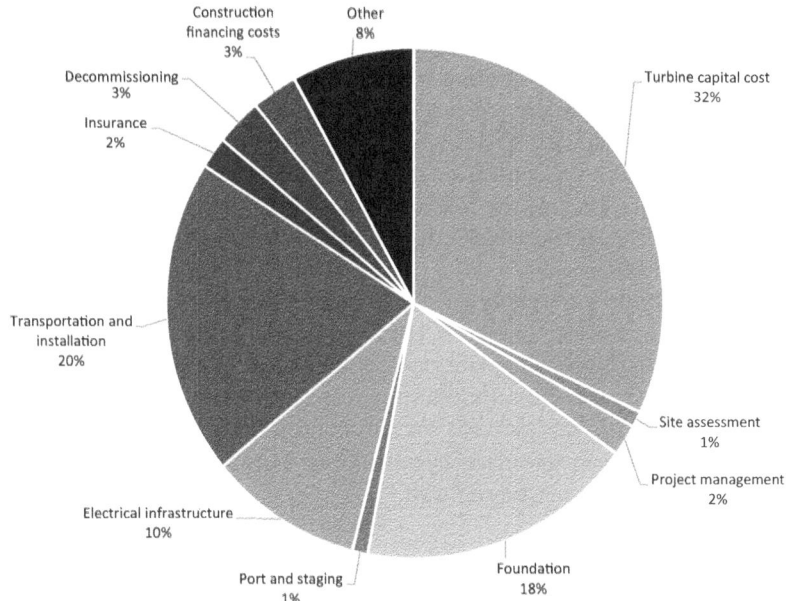

Fig. 3.1   CAPEX breakdown of an offshore wind farm (Source of data: New York Energy Policy Institute 2014)

3.2.1.1 Turbine

The WTG is a purchased product from an original equipment manufacturer (OEM). As such, they are generally offered at a stable price. The unit prices will, of course, vary depending on the specific contract terms based on the number of units sold, the delivery point, and, potentially, existing business relationships with operators and/or the desire to build new business relationships with new operators. There may also be additional costs associated with these units if they are designated as typhoon class machines, as will be necessary in the North American Atlantic seaboard and the East Asian markets. This particular CAPEX cost component also happens to be the most expensive.

Because few turbine manufactures have experience with offshore WTGs, the market has been thus far almost entirely dominated by one WTG supplier, Siemens. As of 2014, Siemens machines accounted for 65% of all turbines installed offshore in Europe (EWEA 2015). This concentrated market structure is, in many ways, an outcome of the temporary exit of turbine manufacturer Vestas from the offshore wind market in 2007. That move effectively left only two offshore wind turbine suppliers in the market: Siemens and Repower (ShrutiShukla 2014). Although this market dominance is slowly easing as non-incumbent competitors secure more commercial orders for competing larger turbine machines, it could be suggested that the dominance of just one player in the market for the past several years has made itself felt in higher prices than would otherwise have been seen, had there been more competition in the market. The emergence of new players into the market suggests that more pricing competition will enter the sector, which could substantially impact the CAPEX for offshore wind farms.

Along with the emergence of new competition, another factor that is having a positive effect on the CAPEX of WTGs is the development of next-generation, high nameplate capacity machines. These larger units are producing more power for a modest increase in material costs and, effectively, no increase in labour costs. As automation becomes more prevalent and new production geomarkets outside of Europe are realised, these labour costs may, in fact, be reduced.

3.2.1.2 Foundation

To date, the relatively shallow water depth of major European commercial leases has allowed for the predominant use of 'monopile' foundation types. These foundations consist of a single member pipe pile and a transition piece. The transition piece is used to ensure verticality of the foundation and has been, historically, grouted to the pipe pile. By 2014, these structures accounted for 79% of all WTG foundations installed in Europe, or 2301 foundations (EWEA 2015). These cylindrical steel structures, suitable for shallower waters and the turbines currently on the market (see Fig. 3.2), generally require less demanding fabrication techniques and installation conditions, for a myriad of reasons. Notably, the steel diameter for the monopiles demanded by the market, generally 3–6 metres, can be 'rolled' in existing fabrication yards in northwest Europe that have traditionally been used for oil and gas. These foundations also lend themselves to automated production as the longitudinal and circumferential welds can be performed by machines. Likewise, installation vessels borrowed from the North Sea oil and gas exploration sector are, in most cases, capable of installing these foundations at the water depths of the projects currently under development.

However, the shift towards larger turbine sizes (>5 MW) with greater loadings, together with a trend towards deeper waters, will all contribute to more strenuous requirements for offshore wind foundations. Not only will more streamlined and serial manufacturing capabilities in fabrication yards be required, but also more purpose-built installation vessels with greater crane hook capacity and larger installation equipment. There have also been a significant number of failures in the European market of the grouted connection between the monopile and the transition piece (Deign 2011). This has led to new solutions involving a flanged driven pipe pile which is then bolted to the transition piece.

Currently, in Europe, new project developments in deeper waters with larger turbines are opting for one of two solutions, the 'XXL' monopile or the jacket-type foundation. The 'XXL' monopiles are being used as there exists a certain amount of comfort in the knowledge that these single pipe members have been used successfully for a large number of projects to date. However, the largest members built as of March 2016 are approaching 8 metres in diameter and weigh 1300 metric tonnes (Garus 2016). The number of facilities that can produce them is quite limited. This leads to a reduction in competitive bids and can drive up costs. Members being designed for future projects are expected to reach 10 metres in diameter and weigh up to 1700 metric tonnes, which is poised to further narrow down the number of capable fabrication facilities in Europe and elsewhere.

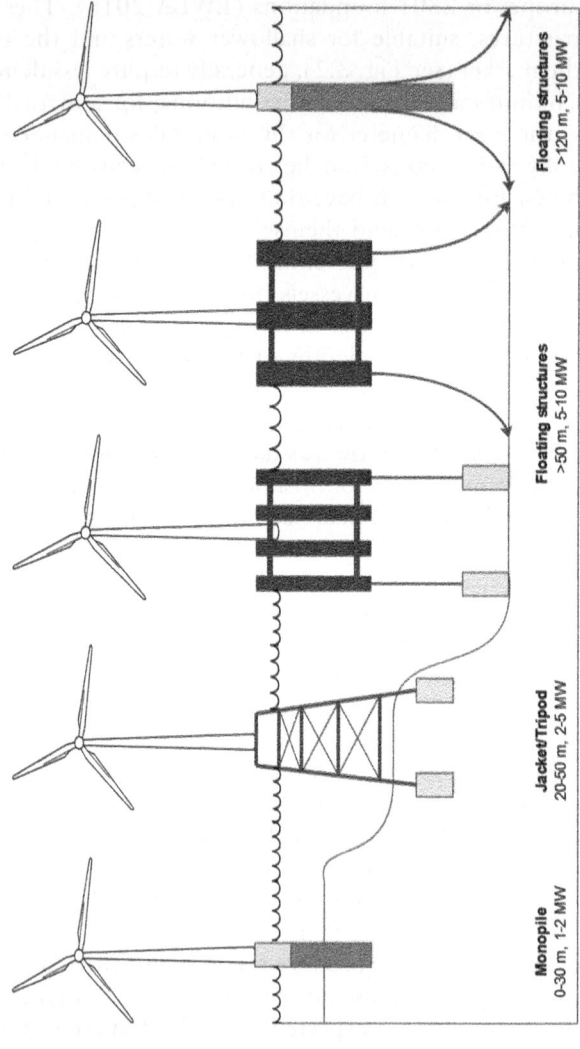

Fig. 3.2 Offshore wind farm foundation types at different depths (Source: Authors)

The jacket-type foundation, a lattice structure of tubular members welded together in various configurations, also has peculiarities that have a negative impact on its cost. The largest of these is that the labour costs associated with these foundations are significantly higher than simple single pile structures. The welding of these braces is, usually, manually performed by skilled craftsmen. In Western Europe, where most of the projects are, these labour rates can be quite high. However, if these projects were to have the foundations fabricated in markets where the labour costs are lower, they would be faced with increased costs to transport these foundations to the project site.

3.2.1.3 Grid Interconnection

The connection of offshore wind power to the mainland grid is costly and the costs further increase as the distance from the coast increases. Unlike the early years of the industry, when installations were close to the coast, most of the current wind farms are relatively distant from the shore, often in excess of 50 km. Being further away from the coast can have several advantages, including reduction in visual impacts and noise emissions, as well as the opportunity to install larger wind turbines with higher throughput. Nonetheless, distance from the shore increases the cost of grid connection considerably, because of the need for special equipment—offshore substations, subsea cables, insulators, switchgears, and protection equipment, among other items—that is compatible with a harsh marine climate. Farm-to-shore grid investment already constitutes 15–25% of the total costs of new projects (Meeus 2015).

At the same time, the design and operation of offshore transmission lines evolve as the need for more optimised grid connection methods increases. To date, wind farms are mostly operating independently of each other. Thus, a point-to-point connection, where each wind farm is directly connected to the mainland grid, has been the common approach. As interconnectivity among wind farms increases, the entire transmission infrastructure needs to be optimised at once, taking into account topology, control, and interoperability of equipment. Some of these changes are already under way as, for example, high-voltage alternating current (HVAC) linking methods are being replaced by a high-voltage direct current (HVDC) link in places where wind farms are situated far from the coast. Figure 3.3 shows, schematically, the offshore grid connection methods.

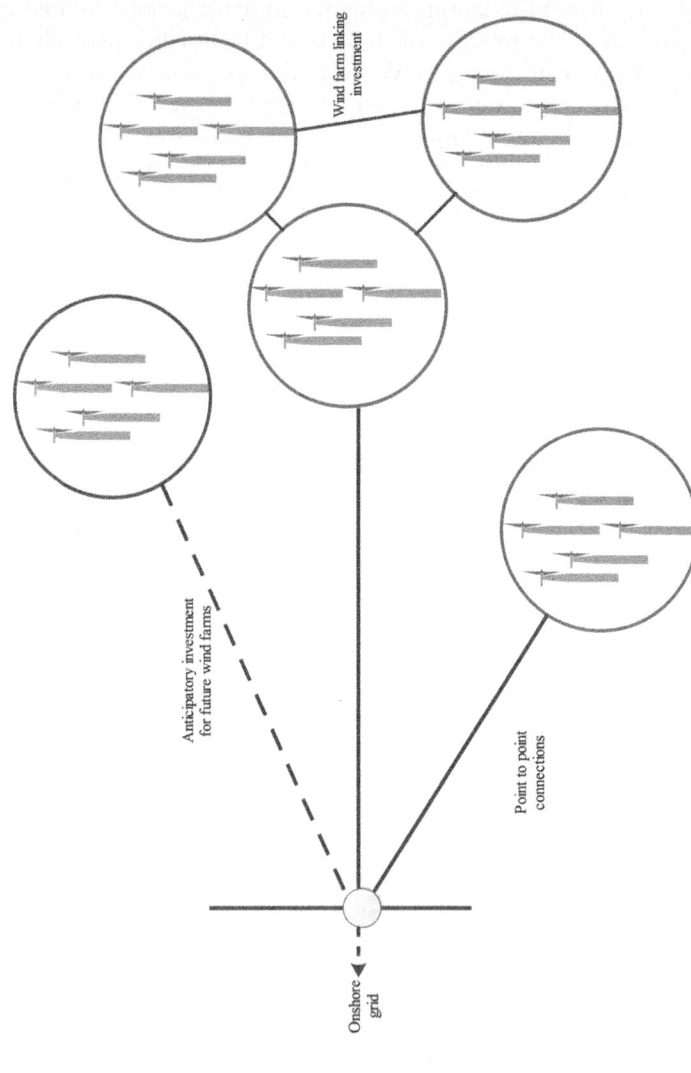

Fig. 3.3 Offshore wind farm grid connection (Source: Authors)

As these projects move further from shore and become larger in scope, it will drive up two types of connectivity costs. The first is the cost of subsea inter-array and export cabling. As projects connect more individual WTGs, they will require more miles of inter-array cabling, and as they move farther from shore, they will require more miles of export cabling. In order to reduce output losses, projects that are further from shore will also likely use HVDC cabling. This material is more expensive per kilometre than HVAC cabling (Williams 2011).

The use of HVDC export cabling also leads to an increase in the costs of the second major electrical infrastructure cost driver, the offshore electrical services platform (ESP). These serve essentially as collector platforms from all of the inter-array cables between the individual WTGs and then transmit the power through the export cable to the shore. These platforms are very large as are their supporting foundations, which are typically supported by jacket-type structures. The larger farms will likely require multiple ESPs all with HVDC conversion equipment.

3.2.1.4 Transport and Installation

Transportation and installation costs consist of the vessel day rates, auxiliary equipment day rates, fuel costs, and mobilisation and demobilisation costs. Vessel day rates vary widely based on myriad of parameters. A non-exhaustive sampling of these parameters includes crane hook load capacity, propulsions systems, deck area, and, perhaps most importantly, company track record. As projects move to deeper water and require heavier offshore lifts for both foundations and WTGs, larger vessels with the crane capacity to make the lifts are required. The relative rarity of these vessels, the laws of supply and demand, and the costs associated with building these larger boats all serve to dramatically increase their day rates.

Auxiliary equipment day rates include all of the varied handling and installation tools that are needed to perform the marine construction works. Notable among these are the pile driving hammers and pile handling tools, the grouting equipment (if the foundation type requires it), and the drilling equipment (if the seabed conditions require it). There is a corollary between the increased day rates of the larger transportation and installation vessels (TIVs) and the auxiliary equipment. As larger cranes are needed to lift larger pieces offshore, so too are larger hammers for driving, larger drills for drilling, and higher output grout pump equipment required.

In addition to this, fuel costs are a commodity cost and vary with the global markets. The increase in costs in this case is strictly related to the fact that larger vessels burn more fuel per knot and the distances being travelled are increasing. This is also the driver for the increase in mobilisation/demobilisation costs. The further afield these vessels and equipment need to travel between project sites, the higher the costs. This is especially true in the emerging markets of North America and East Asia as there exists no consistent project pipeline and the required TIVs do not exist in these areas. This results in exceptionally high mobilisation costs as the specialised TIVs travel from Europe.

The last element of CAPEX, planning and development costs, is fairly well understood in Europe. Although the switch to contract for difference (CfD) schemes in the United Kingdom has generated a new learning curve, in general, the developers in the European market have the planning and permitting costs quite lean as the knowledge base and repeatability are exploited. This is decidedly not the case in North American and East Asia as developers are learning on the job. Further complicating issues in the United States are the variations in policy schemes from state to state. In summation, the planning and development costs are significantly higher in developing markets.

3.2.2   Operating Expenses (OPEX)

The OPEX of an offshore wind farm can be broken into two main categories, operation and maintenance. The operational expenses include ongoing costs such as seabed leasing, insurance, and transmission charges. The maintenance costs include both preventative and corrective maintenance of the offshore assets and the vessels, ports, and personnel costs associated with performing the required maintenance.

3.2.2.1   Operation Costs

Seabed leasing costs vary dramatically from site to site. Leases are generally awarded based on a competitive auction. The more attractive (i.e., stronger wind resource, shallower waters, nearer to shore, favourable seabed conditions, favourable political environment, etc.) a site is, the more potential bidders it is likely to attract, inevitably driving up the lease costs. Particular to the United States, whether the sites are in state waters or federal waters will affect the leasing mechanisms. One would expect, as the new project sites move further afield, there will be less entities with the wherewithal to develop these projects and the margins will likely be less favourable, which could drive these lease costs down somewhat.

Insurance cost should also be expected to increase proportionally with the total value of the marine assets. These larger projects will, as a matter of course, incur larger insurance costs. But, as the total cost per MW is likely to decrease, the insurance costs will likely be cheaper as well when considered on a dollar per MW basis. This trend has the potential to be reversed though as new, unproven technologies (WTGs and foundations) could be seen as posing additional risks to the insurers. Transmission costs must also be considered as the project owners plug in to existing grids and will have to compensate grid owners for the use of this existing infrastructure. This cost would be expected to only increase linearly with the increasing project size and should not create any adverse considerations as the industry develops.

3.2.2.2 Maintenance Costs

Preventative maintenance includes scheduled offshore inspections of all structural, mechanical, and electrical components of the system (GL Garrad Hassan 2013). Structural welded and bolted connections must be inspected on a regular basis to ensure that fatigue cracking (of welds) and loosening (of bolts) are not occurring. Mechanical maintenance includes the scheduled replacement of parts, greasing of bearings, and magnet and turbine visual inspections. Electrical maintenance includes inspection of all transformers, switchgear, and cabling. The costs associated with these works would be expected to decrease on a dollar per MW basis as the same level of effort is required regardless of the nameplate capacity of the WTG.

Corrective maintenance involves the repair of components that have unexpectedly failed (GL Garrad Hassan 2013). Because the emerging markets of North America and East Asia are subject to tropical storm loadings and seismic events, it is likely that these corrective maintenance costs will be more extensive in these geomarkets. It is also quite likely that as new WTG OEMs and, indeed, next-generation WTGs from established OEMs will have initial release issues that could incur higher failure rates resulting in increased costs.

The vessels, ports, and labour costs associated with this maintenance will also vary dramatically from location to location. In the United States, where the bulk of the projects are in the Northeast Atlantic corridor, land is scarce and labour rates are high due to a large union presence and a higher cost of living. These onshore operations and maintenance (O&M) hubs will have to be located near to the actual wind farms, so this will limit the range of port properties that can be considered. A further cost driver is that the O&M vessels in the United States will have to be built within the country due to Jones Act considerations. These US-built vessels will be significantly more expensive than comparable vessels built in the developed East Asian ship-building nations such as China or South Korea. Land is also scarce in the major developing East Asian offshore wind markets such as China, Taiwan, South Korea, and Japan. However, they benefit from cheaper labour rates and the lack of restrictive vessel laws. The one exception to this in this market lies in the difficulty of using Chinese vessels in Taiwan.

In Europe, the benefit of established projects has a trickle-down effect to the O&M costs. Multiple projects can share O&M vessels, onshore base stations, and personnel to spread these costs over a greater area.

3.3   Technical Hurdles

While the offshore wind industry is approaching three decades of existence in the European market, there still exist significant technical hurdles and inefficiencies that have yet to be overcome. The industry developed in a relative bubble and, for reasons unknown, did not draw upon the lessons learned from the longstanding offshore oil and gas industry. There are five significant technical hurdles in the offshore wind sector that, if overcome, can result in significant efficiency for the industry going forward. They are subsea cable reliability, inefficiency in design processes, overcoming restrictions of weather, the inherent intermittency of wind power, and lastly, and perhaps most importantly, commoditisation of foundations (i.e., foundations that can be produced by automation and that are not unique to each individual WTG location).

3.3.1   Subsea Cable Reliability

Subsea cable failures cause more financial losses to the offshore wind sector than any other element. In 2015, these losses totalled more than €60 million and accounted for 77% of total global losses in the sector (Tisheva 2016). Two-thirds of these losses are attributed to contractor error.

Clearly, a large portion of these losses can be mitigated by sharing lessons learned from the previous incidents across the marine construction sector, and it is imperative that these technical liabilities are rectified moving forward. The loss of a subsea cable is not only expensive to repair and replace, it also adds considerable economic losses in the power that isn't being pushed to the grid.

3.3.2   Design Process

Currently, the design processes of offshore substructures and WTG systems rely on iterative processes between the foundation designer and the WTG OEM. That is to say, no single complete system model is being analysed for the expected conditions that the system will see over its 20 plus year life. The responsible parties for each of these components are, to date, without exception, separate entities with separate knowledge bases (hydrodynamic vs. aerodynamic loadings). This results in a design process where the two parties are iterating and resolving all of their complex load results to a single point (normally the tower flange). Because these systems are so dynamically sensitive and fatigue driven, this also means that all of the complex Eigenmodes must be delivered in a large matrix form at this same point. In practical terms, what this means is that the design process, which includes hundreds of thousands of load simulations, must be repeated several times. Typically, a minimum of three cycles is needed to achieve convergence. This adds significant computer costs and engineer labour costs to the design. While the engineering may be a small percentage of the total project cost, it nevertheless remains significant. The ideal solution is a coupled analysis, where the full system is modelled in one software suite and the aerodynamic and hydrodynamic loads are applied simultaneously. This obviates the need for the iterative process and not only saves engineering costs but, more importantly, relieves pressure on the project schedule. This could result in an earlier production of power, thereby greatly increasing the rate of return for the particular project economics. The figurative 'wrench in the works' to this solution is that many OEMs use proprietary software to simulate their WTG controls and aerodynamic loadings and are reluctant to combine their models with existing commercial hydrodynamic software. This appears to be easing as new competition and industry pressure push towards the ideal coupled analysis.

3.3.3   Weather Restrictions

Most offshore wind projects schedule the marine construction for the spring and summer months, when the sea states and wind speeds are relatively mild allowing for more use of cranes (which are limited by wind speeds) and jacking up/down and dynamic vessel to vessel lifts (which are limited by significant wave height). Even during this relatively milder time of year, vessels experience approximately 25% down time in waiting for weather conditions to improve (see Fig. 3.4) (Nielsen 2016).

As noted previously, these vessels have quite expensive day rates and the marine construction contractor must price this risk into his bid. The ability to reduce installation costs would have a dramatic effect on the total CAPEX. This milestone could be readily achieved by developing technologies that allow offshore lifts to occur at higher wind speeds. This will become even more important as WTGs and foundations become larger and have a larger wind drag area. Not only will this development allow more efficient vessel utilisation in the spring and summer, but ideally it would allow for installations in the fall and winter months as well. There are other considerations and restrictions on yearly marine construction works such as marine mammal migration patterns. But, by far, the biggest reason for not working through fall and winter is weather. TIV owners

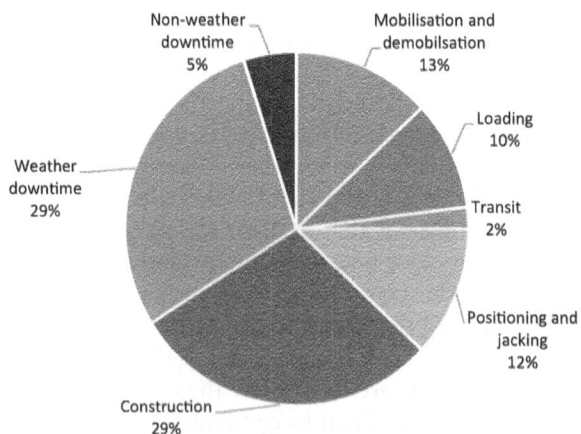

Fig. 3.4 Distribution of vessel costs for a WTG installation (Source of data: New York Energy Policy Institute 2014)

also have to price in these idle months into their active contracts. Greater operability would allow this cost of recovering vessel CAPEX into more projects.

3.3.4 Intermittency

Offshore wind power, naturally, is an intermittent source. Turbines operate in specific ranges of wind speed occurrences. If the wind is blowing too softly, the turbines cannot overcome the inertia of standing still. If the wind blows too strongly, the blades must be feathered out of the wind to avoid overload conditions on the system, both structurally and mechanically. Most of the newer offshore wind projects have capacity factors in the 40%–50% range (i.e., the turbines are producing their rated power between 40% and 50% of the time) (Energy Numbers 2017). The existing grid infrastructures in Europe and in emerging markets are not designed for these high spikes of generation at the windiest times of day. Perhaps more importantly, the traditional energy sources that must remain in the equation because of these intermittency issues have high costs associated with ramping up and down their production to that degree. The operational challenges of the grid in dealing with intermittency issues can be costly as the operator needs to handle voltage variations and fluctuations, reactive power, frequency control, and harmonic disturbances, among other issues. Recently, there has been a push in Europe to move these responsibilities to the wind farm operator and require them to control many of these issues prior to supplying power to the grid. This could have the knock-on effect of driving up the CAPEX costs of the offshore wind farms. Many of these issues could be resolved with innovative storage technologies. Battery technology is drastically improving, in no small part due to the automotive industry's development of electric cars and mass production of lithium-ion storage units such as Powerwall and Powerpack by Tesla. There is also possibility of pumping water to height to develop a potential energy head during times of low usage and using that water to supplement the wind sources during times of high usage. However, the usage of pump storage is constrained by geography. Another innovative method involves the use of power to gas technology whereby surplus electricity is converted to gas and stored in gas storage facilities for later use (to produce electricity or heat).

3.3.5   Serial Production of Foundations

The last major technical hurdle is the commoditisation of foundations. Offshore conditions vary widely across the globe. Every project has different bathymetry, subsea soil conditions, metocean conditions, wind speeds, and turbulence due to the layout of the farm. Some projects have to deal with sea ice or tropical storms or seismic activity. This endless combination and variability of load conditions make it impossible to develop a foundation that is a panacea for the industry. Short of building individual islands offshore to effectively turn the dynamics of the system into that of an onshore turbine, this day will not come. However, the industry can, and indeed must, be smarter about the repeatability of, at the very least, the complex fabrication portions of these foundations. Much in the way that Ransom Olds and Henry Ford were able to drastically reduce the costs of the automobile, so should the offshore wind industry latch onto foundations that lend themselves to this type of construction. In doing so the CAPEX costs of this component should dramatically decrease. As noted earlier in this chapter, the foundation is the second biggest CAPEX cost and is trending towards being the largest as more competition enters the WTG market and the size of turbine increases.

3.4   CONCLUSIONS

The high cost and technical challenges of the offshore wind sector have manifested themselves in the sluggish penetration of offshore wind in the global generation mix. At the moment, both the capital costs and the operating and maintenance costs associated with offshore wind are relatively high. Capital costs of offshore wind include the WTG cost, fabrication of the foundations used to anchor the WTG to the seafloor, electrical infrastructure, offshore installation, and planning and development costs. These figures vary significantly market by market depending on existing infrastructure, the availability of installation vessels, and competition amongst industry participants. The operational expenses include ongoing costs such as seabed leasing, insurance, and transmission charges. The maintenance costs are composed of both preventative and corrective maintenance of the offshore assets and the vessels, ports, and personnel costs associated with performing said maintenance.

Despite nearly three decades of existence in the European market, there still exist significant technical hurdles and inefficiencies that have yet to be overcome. The technical challenges of the offshore wind sector include subsea cable reliability, inefficiency in design processes, restrictions of weather, the inherent intermittency of wind power, and commoditisation of foundations.

However, as it will be discussed in details in Chap. 6 the cost drivers of offshore wind have considerable opportunities to be reduced through lessons learned and efficiency in planning. A developed supply chain coupled with advances in wind turbine technology can help reduce parts of capital and operating costs. Likewise, the technical hurdles all have many dedicated minds that working towards solutions and, ideally, would not be a hindrance to the industry as it develops into new, emerging markets.

REFERENCES

Deign, J. (2011). *Monopile Failures Put Grout in Doubt.* http://newenergyupdate.com/wind-energy-update/monopile-failures-put-grout-doubt

Energy Numbers. (2017). *Capacity Factors at Danish Offshore Wind Farms.* http://energynumbers.info/capacity-factors-at-danish-offshore-wind-farms

EWEA. (2015, January). *The European Offshore Wind Industry—Key Trends and Statistics for 2014.* European Wind Energy Association. www.ewea.org/fileadmin/files/library/publications/statistics/EWEA-European-Offshore-Statistics-2014.pdf. Accessed 21 June.

Garrad Hassan, G. L. (2013). *A Guide to UK Offshore Wind Operations and Maintenance.* Scottish Enterprise and The Crown Estate. https://www.thecrownestate.co.uk/media/5419/ei-km-in-om-om-062013-guide-to-uk-offshore-wind-operations-and-maintenance.pdf

Garus, K. (2016). *EEW Has Produced the Worl's Heaviest Monopile.* www.offshore-windindustry.com/news/eew-produced-worlds-heaviest-monopile

Meeus, L. (2015). Offshore Grids for Renewables: Do We Need a Particular Regulatory Framework? *Economics of Energy & Environmental Policy, 4*(1), 85–95.

MIT. (2015). *Technology Improvement and Emissions Reductions as Mutually Reinforcing Efforts: Observation from the Global Development of Solar and Wind Energy.* Institute for Data, Systems and Society, Massachusetts Institute of Technology. http://trancik.scripts.mit.edu/home/wp-content/uploads/2015/11/Trancik_INDCReport.pdf

New York Energy Policy Institute. (2014). *Offshore Wind Energy and Potential Economic Impacts in Long Island.* www.aertc.org/docs/SBU%20OSW%20Eco%20Dev%20Final%2011-25.pdf. Accessed 10 May 2015.

Nielsen, O. J. W. (2016). *Reducing Weather Downtime in Offshore Wind Installation.* http://www.highwindchallenge.com/2016/06/13/reducing-weather-downtime-in-offshore-wind-turbine-installation/

ShrutiShukla, P. F. (2014, December). *Offshore Wind Policy and Market Assessment: A Global Outlook*. FOWIND, Global Wind Energy Council (GWEC). www.gwec.net/wp-content/uploads/2015/02/FOWIND_offshore_wind_policy_and_market_assessment_15-02-02_LowRes.pdf

Tisheva, P. (2016). *Cable Failures Account for Most of Offshore Wind Losses*. https://renewablesnow.com/news/cable-failures-account-for-most-of-offshore-wind-losses-528959/

Williams, A. (2011). *HVDC vs. HVAC Cables for Offshore Wind*. http://newenergyupdate.com/wind-energy-update/hvdc-vs-hvac-cables-offshore-wind

CHAPTER 4

Renewable Energy Support Policy Design

Abstract As most types of renewable generation technologies are relatively expensive, at the prevailing electricity market prices, government support is crucial to promote their share in the generation mix. Nonetheless, designing a suitable support scheme is not a straightforward task, not least because of the competing and sometimes conflicting objectives of policies. This chapter reviews renewable support models and highlights their main specifications in relation to providing incentive for investment and cost reduction, as well as compatibility with renewable policy design principles. It concludes that the choice of support scheme depends on the characteristics of the electricity market, technology, economic institutions, public acceptance, and the ability of the model to reduce risk to investors and provide efficiency.

Keywords Renewable policy design • Indirect support schemes • Direct support schemes • Policy design trade-off

4.1 INTRODUCTION

Despite enjoying a descending cost trajectory, most types of renewable generation technologies, including offshore wind, are still considered relatively expensive. In many cases, investors cannot rely entirely on the market to recover their capital costs. This means that government support policy is key to creating incentive for investments in renewables. Nonetheless, designing renewable energy support policy is one of the most challenging tasks for decision makers. This is because policy design is a matter of trade-offs amongst a set of principles such as effectiveness (does the scheme incentivise sufficient investment in renewable energy?), efficiency (is it the least cost way of achieving a decarbonisation objective?), and equity (who pays for the cost of renewable policy and what are its implications for the energy market?).

There is always a trade-off between risk and return and investors need to balance this trade-off when making investment decisions. The government support policy plays a huge role in this respect as it influences both the distribution of risk and the level of return to investment. In principle, renewable support strategy therefore needs to consider that renewable investors are faced with various types of risks for which they require either risk mitigation policies or a commensurate return on capital. In the absence of a sufficient level of support, investors will not enter risky markets such as that of renewable energy.

Given the high capital cost of renewables, it is equally important that support policies in place incentivise efficient procurement. Achieving ambitious renewable targets indeed requires policy makers to be concerned about the cost of such targets, as it has direct implications for the government commitment towards renewables. Cost is also a factor that may serve to undermine the legitimacy of renewable support and thus affect the government commitment. Therefore, renewable policy needs to be efficient. This means where introduction of market mechanism is feasible, it should be prioritised over other methods.

It should also be considered whether or not the renewable energy policy may create distortions in the energy market. This is true because whether the cost of renewable policies is borne by the end-users or falls back on the government budget, it will have some implications for the wider energy market. Currently, in many European countries, the cost of renewable subsidies is paid by rate payers through a surcharge on electricity prices. This makes electricity more expensive compared with other energy vectors in end-use markets, and thus it impacts inter-fuel competition (e.g., gas and electricity both compete in the heat sector). In addition, higher electricity prices lower the ability of government to decarbonise heat and transport sectors through electrification. Furthermore, expensive electricity prices encourage self-generation. This comes with a potential negative impact on utility companies' revenues as they may not be able to recover their fixed cost when retail tariffs are not designed accordingly.

In sum, designing a renewable policy is not a straightforward task. In practice, a government can adopt its renewable support strategy based on direct policies or indirect policies (see Fig. 4.1) where each of these broad categories has various models to choose from.

The outline of this chapter is as follows: Sects. 4.2 and 4.3 discuss indirect and direct support mechanisms, respectively, whereas Sect. 4.4 evaluates their relative performance. Finally, we provide some concluding remarks in Sect. 4.5.

4.2   INDIRECT POLICIES

Indirect policies focus on reduction of greenhouse gas (GHG), conventional air pollutants, and dependency on fossil fuels with the aim to shift away from hydrocarbon fuels to renewable energy sources. This means rather than making renewable energy cheaper through subsidies, indirect policies put a cost on sources of pollutants and emissions which are, in essence, the main reason behind the need for having renewables.

The European Union's (EU) commitment to binding GHG reduction under the Kyoto Protocol, for instance, and, subsequently, the EU's 2020 mandates and nationally tailored energy strategies (National Renewable Energy Action Plans [NREAPs]), have all set a foundation for renewable promotion in the member states. In addition, the Paris agreement, which came into force on November 2016, may increase the role of renewable energy sources, particularly in the power generation sector, should national commitments for emission reduction materialise.

The mechanism to control emissions can be through tax policy, regulation, or cap and trade programmes. Tax is specifically very challenging to implement as it can receive public opposition and backlash. Direct regulations, although effective, are often criticised for not being market based. On the contrary, cap and trade programmes are very popular because they are not only market based but also easier to implement from a political perspective.

The EU Emission Trading Scheme (EU ETS) is the largest cap and trade programme in the world, which covers 31 countries (all 28 EU countries plus Iceland, Liechtenstein, and Norway) and approximately 45% of EU GHG emissions. According to this scheme, the companies receive or buy emission allowances which they can trade with each other.

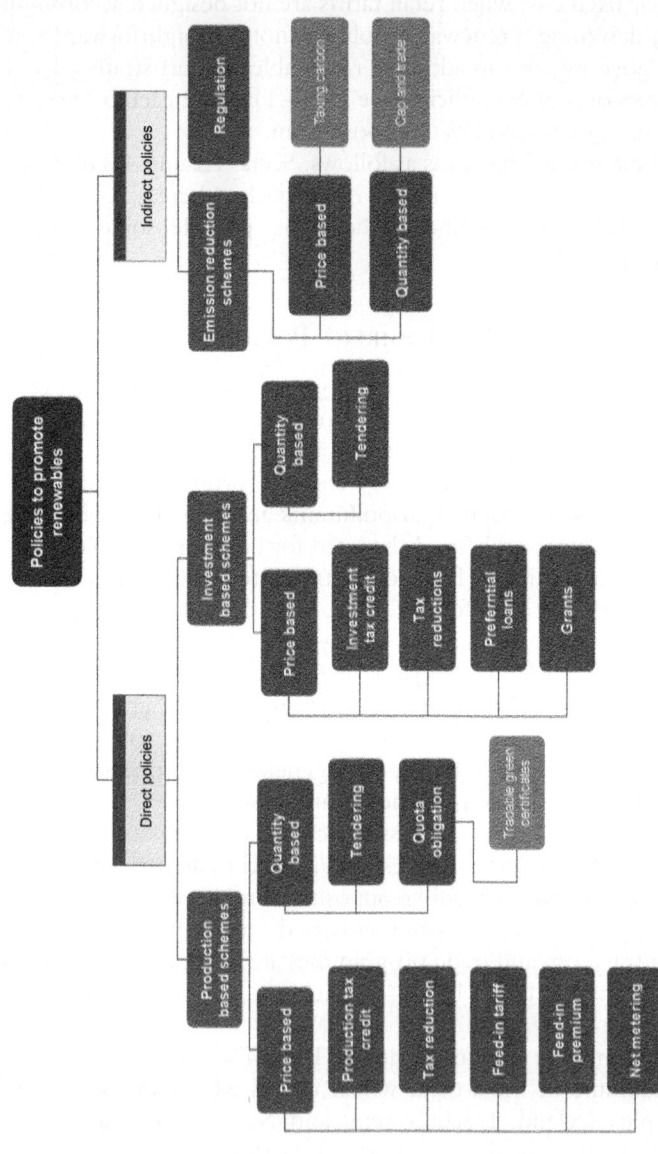

Fig. 4.1 Classification of support policies to promote renewables generation (Source: Authors)

To ensure that emission allowances have a value, their total number in the market is limited. Each year, companies need to submit a sufficient number of allowances to legalise their emission level, otherwise a heavy fine will be imposed upon them.

Indirect schemes such as EU ETS cap and trade programme have several advantages. First, trading provides significant flexibility as it leads to emissions cut where it is least costly. Second, should the scheme be designed adequately, cap and trade can promote investment and innovation in low-carbon technologies.

Nonetheless, EU ETS, as an example of cap and trade, has, in practice, suffered from some design issues. It is sensitive to external factors, and thus a combination of recession, high import of international credits, and industrial lobbying has led to an oversupply of emission allowances in the market. As a result, prices of carbon allowances have slumped to a level that no longer effectively creates any incentive for emission reduction and investment in clean technology. In response to this, the EU has planned to carry out a structural reform on ETS through a combination of short-term and long-term measures. In the short term, auctioning of 900 million additional allowances is postponed until 2019–2020. In the long term, a market stability reserve has been planned to begin operating from 2019. This will be accompanied by a faster reduction of the annual emission cap. The EU member countries can also introduce their own unilateral remedial measures. In the United Kingdom, for example, the government introduced a carbon price floor on power generators via the Climate Change Levy in order to drive up the price of emissions.

4.3 Direct Policies

As opposed to indirect policies, which focus on emission reduction, direct policies aim to promote renewables as a substitute for fossil fuels by providing various forms of financial support (Fischer and Preonas 2010). Therefore, direct policies are more interventionist compared to indirect strategies as they rely heavily on government support. There are various models of direct policies, but overall, they can be categorised as production-based or investment-based schemes. These two broad policies can be further divided into price-based and quantity-based models (Poudineh et al. 2016). In what follows, we discuss these models in more details.

4.3.1   Production-Based Schemes

Production-based schemes incentivise renewables based on the unit of energy they produce and inject into the grid (e.g., $/MWh). This guarantees that renewable facilities operate so it does have a direct effect on emission reduction. In this model, government can choose the price and then let the quantity be determined by the market, or choose the quantity and let the market to discover the price.

There are various forms of price-based production support schemes of which the most important are production tax credit, tax reduction, net metering, feed in tariff (FiT), and feed in premium (FiP). In the production tax credit, the renewable generator will be exempted from tax liabilities partially or totally based on the energy evacuated into the network. Alternatively, renewable generators can be offered a tax reduction according to the energy they have produced. This can include sales, value added, energy, and carbon taxes. Net metering is a scheme under which the owner of a renewable energy facility can meet its own demand through onsite generation and feed into the grid the surplus production (Poudineh et al. 2016). The design of the scheme can vary in different countries, but the main principle is that the net consumption will be metered through a bi-directional metre. The participant in the scheme will pay when the net consumption is positive and will be compensated, financially (at retail tariff often) or in the form of credit for the next billing period, when it is negative.

FiT is a popular renewable support policy scheme in which the owner of a renewable facility is awarded a long-term contract (typically 15–20 years) to feed in energy into the grid. The price of energy is set administratively and it is often higher than retail price. This scheme shields renewable generators from market price volatility and uncertainty for the duration of the contract. FiP, on the other hand, is a policy design whereby the renewable facility is paid a premium in addition to what the generator normally receives from the sale of its energy in the wholesale electricity market. The premium can be fixed, floating, or have a cap and floor.

The production-based support schemes can be quantity oriented. This is usually in the form of tendering or quota obligations. In the tendering model, government determines the quantity of capacity that needs to be procured and then issues a tender in order to set the price (e.g., per MWh) and grant contracts to the least cost developers. The quota obligation, on the other hand, is a policy in which electricity suppliers are mandated to secure a certain percentage of their delivered energy from renewable sources. A Tradable Green Certificate (TGC) (alternative names are renewable energy credit, renewable electricity certificate, tradable renewable certificate, or green tags) is issued by renewable generators for each unit of energy they produce (e.g., MWh). The renewable generators are entitled to the revenue from TGCs, which are traded in a secondary market, on top of the revenue they receive from the sale of their energy in the wholesale market. The electricity suppliers are required to surrender a sufficient number of certificates to comply with their obligation or pay the penalty, which, in practice, is the cap to the price of TGCs.

The support schemes can be combined in order to take advantage of the best features of various models. For instance, FiT can be combined with tendering and contract for difference (CfD) in order to ensure efficiency and take advantage of electricity price movements. Moreover, the government can fix the budget rather than capacity in the procurement process, and thus let the market decide the quantity of installed capacity and price simultaneously. The CfD approach, based on a fixed budget, is adopted as part of the most recent electricity market reform in the United Kingdom to incentivise investment in low-carbon technologies. In this model, a renewable generator sells its energy in the wholesale electricity market, and its revenue will be adjusted upward or downward should the market price be lower or higher than the CfD contract price.

4.3.2   Investment-Based Schemes

Renewable support policy can be investment based rather than production oriented. Similar to the previous case, investment-based policies can be further classified as price-based or quantity-based schemes. Price-based investment-oriented policies include investment tax credit, tax reduction, and preferential loans and grants. On the other hand, quantity-based investment-oriented schemes are in the form of tendering. Investment tax credit and tax reduction mirror their equivalent in the production-oriented schemes with the slight difference that the former are based on capital expenditure rather than the amount of energy produced. Preferential loans are policies through which government covers parts of the interest rate of finance provided to renewable projects by commercial banks or other financial institutions. In the case that the perceived risk is high, government may provide a guarantee to cover a share of outstanding loan principal with the aim to reduce the cost of capital to the investors.

In the investment-oriented quantity-based schemes, government specifies the capacity of procurement and runs a tender to set the price of capacity (per kW or MW) and identify the least cost investors. The government then subsidises the investment (per kW or MW) or partially finances the project in return for an equity ownership.

4.4   EVALUATION AND DISCUSSION

From an economic perspective, indirect policies are more efficient if they are based on pricing carbon emissions rather than direct regulation. This is because they address the prevailing externality which is emissions rather than introducing extra distortion in the market (in the form of renewable subsidies). The carbon pricing can create incentives for investment and innovation in low-carbon generation, with minimum government intervention. In this way, the market, rather than the government, decides which technology needs to be deployed and what is the appropriate mix of technology in a decarbonised economy. Therefore, in theory, indirect policies, based on carbon pricing, are both effective and efficient.

However, in practice, the carbon price at which a shift away from fossil fuels is triggered can be so high that it would make it politically impossible for most governments to impose a tax on carbon. Furthermore, cap and trade programmes such as EU ETS, which are designed to bypass the challenge of direct taxation, and therefore rely on the shadow price of carbon, are often sensitive to various external factors. This means they can become irrelevant to the market if they are not designed to be robust enough to withstand change in market conditions. Thus, in the absence of an effective indirect policy to promote renewable generations, direct polices are the method of choice. However, both direct and indirect policies can co-exist as this is the case under the current renewable policy in the EU region.

Direct polices are generally effective in realising renewable targets; however, designing a suitable direct support policy requires giving careful considerations to a set of criteria such as:

- Reduction of risk to investors
- Provision of efficiency
- Compatibility with the design of electricity market
- Suitability for the scale of the project and stage of market maturity
- Compatibility with the economic/legal institutions and robustness to withstand change in the market condition

As investors are exposed to various risks, a support policy is needed to mitigate these risks, because they have direct impact on the capital cost of renewable projects. For instance, among various direct policies, FiT is a scheme that provides full protections against market price risks through offering a long-term revenue guarantee contract between developer and government. It is also often accompanied with a Power Purchase Agreement (PPA) between two parties to remove volume risk. This brings certainty in renewable generation revenue and ensures investors of profitability of their investment. In fact, this feature applies to all schemes that offer a long-term PPA irrespective of the way that price is determined (administrative or auction). However, other schemes such as FiP and TGC do not provide full protection against economic risks as investors can be exposed to market price fluctuations.

At the same time, support policy needs to be efficient. Schemes such as FiT, in which price is set administratively, can be inefficient if the price is set too high or too low. Additionally, inefficiency of administrative schemes makes them susceptible to political intervention through price adjustment or policy change (e.g., abandoning support). Something that sends the signal of policy instability is much to the detriment of the investment climate. On the other hand, competitive schemes ensure only the least cost investors are awarded the contract. Therefore, competitive schemes, when feasible, should be preferred over non-competitive approaches.

In addition, the renewable support policy also needs to be compatible with the structure of the electricity market; otherwise, the market will not produce an efficient price. This is specifically a challenge in liberalised electricity markets in which power plants dispatch and market prices are determined according to merit order and the system marginal price, respectively. Thus, when regulation gives priority dispatch to renewables, this can lower the market price because expensive plants will fall out of the merit order as their output will not be needed. In places that priority dispatch does not exist, renewable generators outbid conventional generators because they have zero marginal cost and guaranteed out-of-market payments. The outcome, again, is depressed electricity prices. The distortion resulting from renewable policy is currently a major challenge of European energy-only electricity markets[1] which has resulted in conventional generators not being able to recover their capital costs. This has triggered various government interventions including the introduction of capacity markets and change in the forms of renewable support policies. The EU commission recently advocated removal of priority dispatch along with the introduction of FiP as ways to deal with distortions of the energy-only market.

In the non-liberalised markets, where there is a single buyer, the interaction of market and renewable support policy is less of an issue because popular schemes such as FiT and auction can be easily implemented. However, other schemes such as FiP and TGC, which rely partially on market price, require a liberalised market.

One important consideration is that the renewable support scheme needs to be suitable for the scale of projects and stage of market maturity. For example, introduction of investment/production tax credits to incentivise rooftop solar PV may not be effective because households often do not have such tax liabilities to offset. Generally, for the large projects and in mature markets, where there are a significant number of potential investors, exploiting the power of competition, through auction, can reduce the costs and incentivise innovation. In contrast, in the nascent markets, where there are not many potential developers, or when the project size is very small (as in the case of household solar PVs), competitive schemes may not be feasible and may have adverse effects (such as reducing the gain from economy of scale).

Another important point is that a renewable support scheme needs to be compatible with the legal and economic institutions in the jurisdiction it is applied. For example, in places that tax institution is weak or non-existent, incentive schemes such as production or investment tax credit will not be suitable as they hardly create any incentive for renewable investment. This is the case, for example, in the resource-rich countries of the Middle East and North Africa (MENA) where governments do not tax citizens and businesses and instead pay various forms of subsidies (e.g., reduced energy prices).

The renewable support scheme also needs to be robust enough to withstand change in market conditions, as this will provide certainty over the future revenue stream of the project. For instance, schemes such as TGC are not robust as the certificate prices can decline if the installation of new renewable capacity grows fast and the market becomes oversupplied. In this case, those renewable generators which are already operating in the market may not be able to recover their costs. Furthermore, when support scheme exposes generators to market uncertainties, investors need to pay risk premium in order to finance their project something which increases their costs of capital (Krupa and Poudineh 2017). One of the reasons that the UK government abandoned Renewable Obligation (RO) in favour of CfD contracts was that RO increases the costs of decarbonisation by unnecessarily subjecting renewable generators to market uncertainties.

The main message of the above discussion is that the choice of renewable support scheme depends on various factors such as the characteristics of the electricity market and requires giving careful considerations to trade-offs involved. The support scheme can be technology neutral or technology specific. It can focus on the supply side of the market (such as FiT and tendering) or demand side (e.g., TGC). As previously stated, from a policy point of view, a support scheme needs to provide adequate incentive for renewable investment (i.e., be effective) and cost reduction (i.e., be efficient) and operate in harmony with the market system in which it is implemented. Indirect policies are often more efficient, but the challenge is that their implementation, in a way that creates sufficient incentive for investment, is not straightforward. Direct policies are usually more effective, but they rely on heavy government intervention in the market and not all of them are efficient.

Finally, a consideration from a policy point of view is who should pay for the cost of renewable support policy and what is the implication of the adopted model for the wider energy market. Indirect policies have the advantage that the cost of policy will be transferred to the producers of the externality (i.e., fossil fuel power plants) and thus are more conducive to an equitable solution. In addition, indirect policies are more compatible with the wider objective of decarbonising the economy. With regard to direct policies, however, there are two ways to recover the cost of subsidising renewables: from rate payers (electricity users) or from tax payers (general public). In most parts of the world (including the EU region), the cost of renewable policies is imposed on rate payers in the form of a surcharge on electricity prices.

The rate payer approach for covering the cost of subsidies can have two effects on energy markets. First, it raises the cost of electricity and creates incentive for self-generation (i.e., load defection) and leaving the grid (i.e., grid defection), with remaining on-grid customers needing to shoulder the fixed cost of the system. As more electricity users leave the network, the cost will be divided over fewer users and this further increases the unit cost of electricity and further encourages self-generation. This phenomenon, which is known as 'death spiral', has adversely impacted the business model of traditional utilities. It also runs the risk of creating an inequitable two-tier system in the sense that those low-income households that cannot afford self-generation will need to shoulder a higher share of power system costs. Second, the more ambitious the renewable target is, the higher the cost of policy and, consequently, the surcharge on electricity prices. This makes electricity more expensive compared with other energy vectors such as gas. This is problematic for decarbonisation of the heat and transport sectors through electrification, which is currently the default strategy for decarbonisation in the EU region. Therefore, the fiscal aspect of renewable support policy needs to be not only equitable but also compatible with the wider objective of decarbonisation of the economy.

4.5   Conclusions

Given that most types of renewable generation technologies are still expensive, at the current market prices, government support is needed if the share of alternative energy is to increase in the generation mix. However, designing an appropriate support scheme is a challenging task, not least because policy makers need to consider carefully the various trade-offs involved and make an informed decision.

Overall, there are two main approaches for supporting renewable generation: indirect and direct policies. Indirect policies have the advantage that they address the main externality that has necessitated renewable generation (carbon emission) and thus are more conducive to an efficient market solution. In this approach, the government only needs to correctly price carbon emissions and let the market do its job with respect of providing incentive for investment and innovation in low-carbon technologies. Carbon emission can have a real price, by means of taxation, or a shadow price through, for example, a cap and trade programme. However, in practice, taxation is challenging as it is likely to face public opposition. Likewise, designing an effective cap and trade system is not straightforward as the experience with EU ETS shows that such a scheme can be susceptible to various external factors. Accounting for the multitude of factors at the design stage may not be possible.

Unlike indirect methods, direct schemes focus on providing financial support for renewables so that they substitute fossil fuels. This is often achieved through offering subsidies to renewable generators. Although offering subsidies to renewables can be easier compared with taxing emissions, their implementation requires giving careful consideration to the trade-offs involved. From a policy point of view, not only a renewable subsidy scheme needs to be effective, efficient, and equitable, but also it must be compatible with the design of the electricity market, robust enough to withstand changes in market conditions, compatible with the economic/legal institutions, and suitable for the scale of the project and stage of market maturity. In addition, it must not have negative side effects on the wider energy market and should refrain from arousing public opposition (such as tax-based policies). In practice, designing a policy that satisfies all these criteria can be a formidable task.

NOTES

1. Energy-only electricity market, the prevailing model of the EU region, is a form of electricity market in which remuneration to generators is based on only the energy component (so there is no capacity payment). In this model, price spikes at time of scarcity are expected to create sufficient incentive for investment.

REFERENCES

Fischer, C., & Preonas, L. (2010). Combining Policies for Renewable Energy: Is the Whole Less Than the Sum of Its Parts? *International Review of Environmental and Resource Economics, 4,* 51–92.

Krupa, J., & Poudineh, R. (2017). *Financing Renewable Electricity in the Resource-Rich Countries of the Middle East and North Africa: A Review* (Working paper no. EL 22). Oxford Institute for Energy Studies.

Poudineh, R., Sen, A., & Fattouh, B. (2016). *Advancing Renewable Energy in Resource-Rich Economies of the MENA* (Working paper no. MEP 15). Oxford Institute for Energy Studies.

CHAPTER 5

Current Support Policies to Promote Offshore Wind Power

Abstract Given the cost and risk involved, the growth and development of the offshore wind industry are still reliant on government subsidies. This chapter reviews the support schemes implemented in the main off-shore wind markets of Asia, Europe, and the United States, and evaluates their performance. The examination of offshore wind markets shows that the design of an effective support scheme depends, to a large extent, on the specifications of the market and the degree of technology maturity. The results of our analysis in this chapter provide insights on the ways in which support schemes for the offshore wind industry can be designed and implemented.

Keywords Offshore wind support policy • European offshore wind market • Asian offshore wind market • American offshore wind market

5.1 Introduction

Both indirect and direct policy mechanisms have been employed in an effort to curb emissions and incentivise the growth of renewables. Theoretically, indirect policies (which seek to put a cost on sources of pollutants and emissions) are more conducive to an efficient solution. In practice, however, the current indirect policies have not proven effective to incentivise investment in low-carbon technologies on their own, particularly for highly capital-intensive industries. As a case in point, the historical price of carbon under EU ETS (which is the largest carbon market in the world) has been too low (see Fig. 5.1) such that it has not even incentivised a coal to gas switch, let alone a switch to offshore wind power. This means that, until the price of electricity internalises the cost of carbon emissions, direct support policies in the form of subsidies and long-term revenue guarantees are crucial for investors.

A number of direct policy mechanisms have been implemented around the world with the goal of providing support and incentives for the development of offshore wind energy. In this chapter, we review these support mechanisms and evaluate their performance. Section 5.2 describes the direct policy support mechanisms currently in practice. Section 5.3 provides specific examples of the policies that are implemented in global offshore wind markets, as well as a detailed table of the direct policies implemented in several offshore wind markets. In Sect. 5.4 we evaluate these support schemes and discuss their effectiveness based on the stage of market and technology maturity. Finally, Sect. 5.5 concludes.

5.2   DIRECT POLICY MECHANISMS TO SUPPORT OFFSHORE WIND

Direct policies aim to promote renewables as a substitute for fossil fuels by providing various forms of technological and market-specific support. Direct policies can be quantity based, in which government tenders for technology-specific (or technology-neutral) capacity create a market demand for electricity supplied by specific forms of technology. This typically occurs in the form of Renewable Obligations (ROs) (also called Renewable Portfolio Standards [RPSs]) that impose quotas on power suppliers to procure specific volumes of renewable energy from specific generation technologies. Examples of this policy (as described further in Sect. 5.3) include the previously implemented UK RO scheme and, more recently, the US State of Massachusetts, which has carved out a mandate that utilities procure 1600 MW of offshore wind.

In addition to quantity-based schemes, governments may offer production-based incentives that seek to support investments in offshore wind projects by supporting investor confidence in the plant economics (e.g., by providing full revenue insurance to investors). This is typically achieved by governments by providing long-term revenue certainty in the form of 15- to 20-year fixed feed-in tariff (FiT) contracts that guarantee a

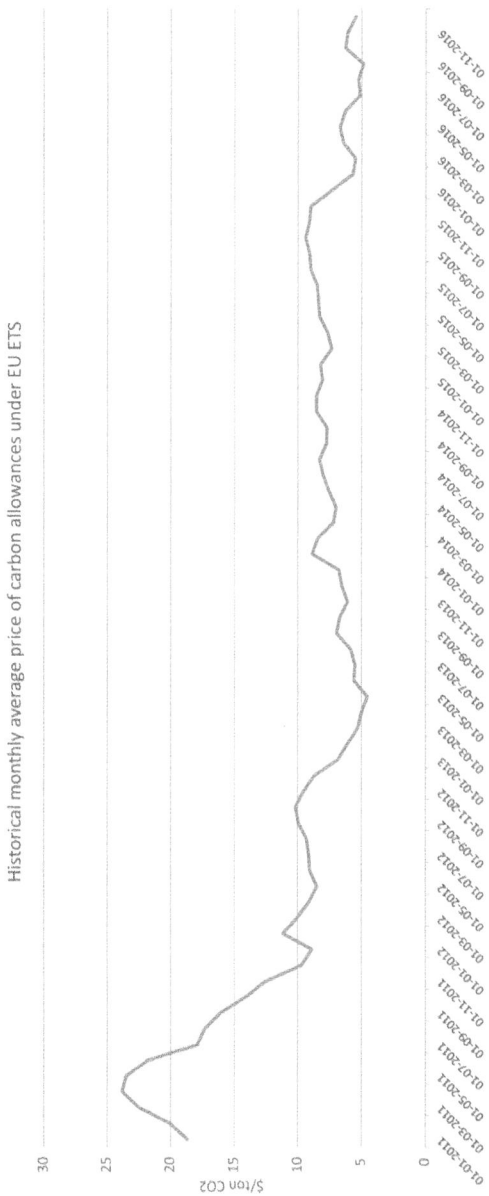

Fig. 5.1 Historical monthly average of daily prices of CO_2 allowances under EU ETS (Source of data: authors collected from Argus European Emissions Markets)

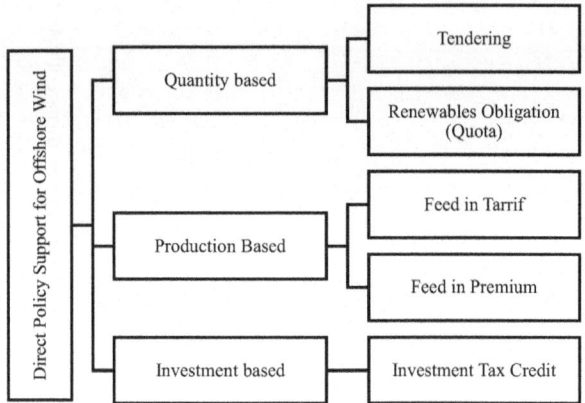

Fig. 5.2 The main direct policy support mechanisms adopted for offshore wind (Source: Authors)

certain price for power sold. Alternatively, a 'feed-in premium' (FiP) model provides a 'top up' payment on top of the wholesale electricity price. In some cases, the value of the FiP is derived from a market for tradable renewable certificates, while in other markets, the premium may be in the form of a cash payment resulting from a reverse-auction subsidy bid.

While direct policy mechanisms to support offshore wind may be similar in structure across markets, there are important differences in specific policy design due to differences in market structure (liberalised vs. centralised), the maturity of the industry in the market, as well as maturity of the technology. Figure 5.2 presents the major direct policy mechanisms in place for offshore wind. This is followed in Sect. 5.3 by an overview of offshore wind support schemes currently implemented in global offshore wind markets.

5.3   Offshore Wind Support Schemes in Global Markets

5.3.1   Europe

The past five years have seen a substantial transformation in direct policy support mechanisms for offshore wind in Europe. Primary drivers of these policy transitions have been, on the one hand, fatigue over the cost of policy support for offshore wind (as well as other renewable energy technologies) in the key markets such as the UK and Germany, and on the other hand, a desire to shift focus from establishing offshore wind capacity to promoting efficiencies and cost-competitiveness with alternative generation sources. Charting these transitions in policy provides insight into the lifecycle of direct policy support schemes for offshore wind, as well as a foundation for understanding the intentions of several options of direct policy support schemes.

5.3.1.1 United Kingdom

The UK is the largest offshore wind market of the world. As of 2016, the total installed capacity of offshore wind power reached more than 5 GW in that nation. The UK support model for renewable energies, and, by extension, offshore wind, has gone under major reforms over time, which have been based largely on the pace of development of offshore wind technology as well as the government's budgetary considerations for supporting renewable energy technologies.

Dating back to 2002, the UK had applied a technology-neutral renewable support model based on RO and tradable renewable obligation certificates (ROCs) (alternative name is Tradable Green Certificate [TGC]). In 2009, however, the government adopted a technology banding approach in which a multiplier was applied to ROCs depending on the band in which renewable technology is placed. For example, for offshore wind power (along with biogas, geothermal, PV, and ocean energy) a multiplier of 2 ROC/MWh was applicable. The idea behind the technology banding approach was to strengthen the incentive for investment in costlier (i.e., less established) technologies. However, since 2014, and as part of the UK electricity market reform, the government has replaced the entire ROC approach with a contract for difference (CfD) model, whereby the project's developer essentially submits a competitive low bid to the government for a tariff price and winners of the reverse-style auction are entitled to a 'top up' payment.

The top up payments, which are calculated as the difference between 'strike price' and hourly electricity market price, are provided to generators when the market price is below the strike price of the CfD. When the market price is higher than the strike price, then generators must return the difference. In this way, the CfD mechanism is intended to provide certainty and stability of revenues to generators by reducing their exposure to volatile wholesale prices (an advantage over the ROC mechanism) whilst at the same time protecting consumers from paying higher tariffs when electricity prices are high.

The CfD acts as a 15-year contract (the amount of full load hours is not fixed) between a low-carbon electricity generator and the government. As well as ensuring the lowest possible price tag for offshore wind support, the guaranteed long-term revenue also seeks to open the door to new forms of project finance and lower the overall financing risk premium. This new renewable policy framework is largely a response to fatigue over high subsidy prices for offshore wind and ratepayer burden. Under the new CfD scheme, a fixed budget is allocated to different carbon-free generation technology pots (Table 5.1), and generators within that pot must compete for, or bid on, the limited number of contracts available for that technology class 'pot'. Under the new framework, 'pot 1' includes established technologies such as land-based wind, solar PV, and hydro. Offshore wind is considered a less established technology, categorised under 'pot 2', along with such technologies as tidal stream, wave, and advanced conversion technologies.

Thus far, the first and second arounds of CfD auctions heave been held and their results are available. The first CfD auction round commenced in October 2014 and saw CfD contracts awarded to both Mainstream Renewable Power's Neart na Gaoithe (448 MW) and Scottish Power Renewables (UK) Limited East AngliaOne (714 MW) projects, for 'strike prices' of £114.39/MWh and £119.89/MWh, respectively (UK Government 2015) and a total capacity of 1162 MW. Although the policy mechanism itself demonstrated drastic price reductions and convergence on the government's established levelised cost of energy (LCOE) price target of £100/MWh by 2020, a lack of clarity over the timing of future auction rounds, at that time, was perceived to threaten investments in the sector.

The long-delayed second CfD round, which awarded projects in September 2017, saw strike prices fall substantially from the previous round. Of the three winners of the 2017 CfD round, the 860 MW Triton Knoll project is proposed to be delivered at a strike price of £74.75/MWh and is due for commissioning in 2021–22. The 1,368 MW Hornsea Two project and the 950 MW Moray East Offshore wind farm, both slated to be commissioned in 2022 –23, each cleared at strike prices of £57.50/MWh. The 2017 UK CfD auction round has indeed shown encouraging price reductions that mirror improvements in project economics elsewhere in Europe. However, the ultimate proof of whether the projects can be delivered at

Table 5.1 UK CfD budget—total renewables energy support payments by delivery year (£million)

Year	2015–2016	2016–2017	2017–2018	2018–2019	2019–2020	2020–2021	2021–2022	2022–2023
CfD budget	50	220	325	325	325	325	290	290
Pot 1 (established)	50	65	65	65	65	65	–	–
Pot 2 (less established)	–	155	260	260	260	260	290	290

Source: DECC (2015) and BEIS (2016)

these prices will not come until these projects approach Final Investment Decisions, presumably around 2019–20. Thus, although the second CfD round appeared successful, there are concerns about the viability of cleared projects and their actual delivery. This issue is explored more in Chap. 9.

5.3.1.2 Germany

Germany's direct support scheme for offshore wind has also undergone significant transformation in recent years. The support scheme for offshore wind farms is currently an FiP model. The premium is calculated as the difference between market price and the maximum guaranteed price which has, traditionally, been set administratively. Under the amendment of the German Renewable Energy Act (EEG) in 2014, offshore wind farm developers could choose between two models (BMWi 2015). The basic model provides a guaranteed maximum payment of €154/MWh for the first 12 years of operation, which can be extended for all wind farms that are located beyond 12 nautical miles from shore and at depths greater than 20 metres. After this period, the support will decline to the basic level of €39/MWh. The second model, however, offers an acceleration scheme in which financial support of €194/MWh is provided for only eight years, which might be extended depending on the distance from the coast and water depth of installations. After this phase, the basic model of €39/MWh will apply. The front loaded (also called compression) model, which offers the higher initial payment for a shorter period, is designed to allow investors to pay off their finance debt more quickly, and thus seeks to reduce the cost of project financing. Initially, the acceleration model was only applicable to wind farms operating on or before 1 January 2018, but was later extended to the end of 2019 in the amended 2014 EEG. The duration of the support scheme is 20 years without a cap on the number of full load hours that are eligible for subsidies. Also, in both models, during the period of low tariff, offshore generators can retain market price should it exceed the guaranteed price.

On 7 July 2016, the German Upper House approved amendments to a number of acts including the EEG (Watson Farley & Williams 2016). The amendments to the EEG, which come into effect in 2017 (thus called EEG2017), state that renewable energy projects shall be subject to auctions and can no longer be remunerated through administrative prices. A separate Offshore Wind Act (WindSeeG) has also been approved that specifies the auction model for offshore wind power. The first round of this new auction model was implemented in March 2017. However, the EEG2014 was in force until 31 December 2016, and its transitional period (to auction system) is unchanged—meaning those offshore wind generators which received a firm grid connection confirmation before 1 January 2017 and are due to be commissioned before 31 December 2020 are exempted from auction. A transitional period is also applicable to offshore wind generators that are going to be installed between 2021 and 2025. The new auction system model will be fully applied to all offshore wind farms from 2026 onward.

On 3 April 2017, the German offshore wind auction cleared with results that exceeded the expectations of analysts. EnBW (Germany utility company) and DONG (Danish Energy Company) won the auction to deliver (in years 2024 and 2025) four projects- three of which, with a total capacity of 1380 MW, without any subsidy. These projects, if realised, will be the first offshore wind farms in the world that are developed entirely based on electricity market prices and without any subsidy. However, there are concerns that whether such low prices are indication of real costs cut in the industry or they are related to other factors such as auction design. This issue will be examined in Chap. 9.

5.3.1.3 The Netherlands

The Netherlands has a target of 4.45 GW of offshore wind to be deployed by 2023 as parts of its renewable strategy. For this to happen, it has selected three main offshore wind zones to develop: the 1.4 GW Borssele zone, the 1.4 GW South Holland coast, and the 700 MW North Holland coast, with a total capacity of 3.5 GW (NEA 2015).

To develop the sites, the Dutch support scheme is a tender-based FiP that awards a subsidy and consent to build and operate an offshore wind farm to the lowest cost developer for each specified zone. Uniquely, the government provides a 'one-stop shop' model for these lease zones, tendering fully consented sites to reduce upfront development costs. Offshore wind generators sell their electricity in the market and receive a premium as the difference between the maximum guaranteed price (price required to cover the cost of generators) and corrected market price (which is determined annually as the average electricity market price corrected for imbalance and profile risk). The maximum guaranteed price is determined through tender and is capped in order to ensure cost reduction. The cap price will decline to €100/MWh in 2019 as seen in Table 5.2. The maximum price is not adjusted for inflation during the term of the contract, and the maximum amount of production that is eligible for subsidies is capped based on a certain amount of full load hours, which varies by project. The annual under and over production can be banked.

Table 5.2 Timetable for tenders of Dutch offshore wind farms

Year	Zone	Capacity (MW)	Price cap (€/ MWh)	Winning bid (€/MWh)
2015	Borssele (sites I and II)	700	124	€72.70
2016	Borssele (sites III and IV)	700	119.75	€54.50
2017	South Holland coast	700	107.5	TBD
2018	South Holland coast	700	103.25	TBD
2019	North Holland coast	700	100	TBD

Source: NEA (2015) and authors' own compilation

To date, this model has shown great success in bringing competitive bids and lowering the cost of subsidies. The permit and subsidy to develop wind farms at Borssele sites I and II were awarded, through a competitive process, to Danish Oil and Natural Gas (DONG) Energy in summer 2016 with an average strike price of €72.7/MWh, excluding transmission costs. This price is considerably lower than the price cap, even after taking into consideration that the government paid for the key elements of project development and has taken the risk of grid connection (which is estimated to add a modest cost of €14/MWh had it been taken into account for bidding). The contract was awarded for 15 years, after which time the developer will receive market price.

Subsequent to this, the tendering round for Borssele sites III and IV saw a consortium of Shell, Van Oord, Eneco, and Diamond Generation Europe (a wholly owned subsidiary of Mitsubishi Corporation) win the tender at a strike contract price of €54.5/MWh in December 2016, which was not only far less than the cap of €119.75/MWh, but also among the lowest prices for offshore wind in the world. These particular auctions, in addition to the Kriegers Flak subsidy auction in Denmark the same year (see Sect. 5.3.1.4), ushered in a new era of cost efficiencies for offshore wind farms.

5.3.1.4 Denmark

Despite its small size, Denmark has been one of the most successful countries in the development of offshore wind power. It was the first country in the world to install a wind farm at sea (Vindeby) in 1991 and has recently demonstrated the lowest bid prices in the world, at less than €50.0/MWh. As of year-end 2016, the country had approximately 1271 MW of offshore wind energy capacity installed and has one of the highest rates of wind penetration in the generation mix in Europe.

Denmark supports offshore wind installations through a premium tariff that is the difference between guaranteed price and market price. Offshore wind developers are required to actively participate in the wholesale electricity market and sell their energy so they are fully integrated within the existing liberalised electricity market. The FiP varies according to the market price and the guaranteed maximum price, which is determined through an auction. When the market price exceeds that of the guaranteed price, no payment is made to the wind farm operator; rather, the upside gain is used to offset the payments in the following period. Additionally, in recent auctions, offshore wind farms are not eligible for support when the market price is negative. This is done to prevent production when the system generation exceeds demand and there is little flexibility to absorb surplus power.

Notably, in November 2016, the contract to build and operate the 600 MW Kriegers Flak offshore wind farm in the Danish part of the Baltic Sea was won by Sweden's Vattenfall with a strike price of €49.9/MWh (37.2 Danish øre/kWh), which is 58% below the original cap of €120/MWh (offshoreWIND.biz 2016). As of 2016, this is the lowest price for offshore in the world and will also be the largest offshore wind farm in Denmark and Baltic Sea.

The record low cost of the Kriegers Flak offshore wind farm can be attributed to a number of factors that have implications for market and policy design. First, Denmark is a mature market for offshore wind in which an increasing number of well-established players compete. The country is home to the world's largest offshore wind developer, DONG Energy, in addition to two of the largest offshore wind turbine original equipment manufacturers (OEMs), Siemens and MHI Vestas, major offshore fabrication yards and experienced engineering companies. This inherently places a downward competitive pressure on the clearing price for market entrants vying to compete for offshore wind farms. Second, the cost of offshore wind technology has also been on a descending cost trajectory. On a per MW basis, new larger advanced turbines require less foundations, less cabling, less installation work, and less maintenance compared to previous generation turbines. This all supports lower bid prices for tenders. As such, Denmark, the first market for offshore wind farms in the world, continues to help pave the way for the offshore wind industry.

5.3.1.5 *Belgium*

The offshore wind sector is expected to significantly contribute to the renewable energy target of Belgium. As of 2016, the installed capacity of offshore wind reached 712 MW in this country.

The support scheme for offshore generation in Belgium is based on a green certificate which is bought by the grid operator, Elia. The value of the certificate is the difference between the guaranteed set price (determined based on the average LCOE of offshore projects) and the corrected market price which, in practice, makes the Belgian support model a form of FiP. The corrected electricity price is the average market price for a year reduced by 10% to compensate for the fact that investors are encouraged by lenders to enter into long-term contracts (given the difference between the price of long-term forward contracts and spot market). The offshore wind operator can receive a maximum price, which is currently set at €138/MWh. For offshore installations after 30 June 2017 and before 1 January 2021, the LCOE can be revised every three years if there is evidence of cost efficiency in technology (TKI Wind OP ZEE 2015).

In Belgium, the duration of support is 20 years and the subsidised electricity output is not subject to a cap on full load hours. After this period, the offshore wind developer will receive the market price. In addition, the guaranteed price is not adjusted for inflation and, if the cost of grid connection is paid by the owner of the wind farm, the maximum guaranteed price will rise to €150/MWh. Also, if there is a surplus of production in the market, the price of the certificate will drop to zero for a maximum of 72 hours a year.

5.3.2   Offshore Wind Market Outside Europe

Over the past decade, several Asian markets, motivated by energy security concerns, growing urban populations, land scarcity, and an abrupt U-turn from nuclear power, have also explored offshore wind. However, given the fact that the infrastructure is less developed in many Asian markets and the electricity market designs are different than those typically experienced in the West, policies that have been borrowed for offshore wind support from Europe have also been adapted to suit the market context.

The offshore wind market is also getting off the ground in America. However, there are still significant uncertainties that exist. The following subsection reviews the offshore wind policy frameworks and progress in key Asian markets as well as the United States.

5.3.2.1   China

The offshore wind industry is considered to be important in China because of increasing electricity demand, proximity of wind resources to local demand centres, and efforts to decarbonise power generation and improve air quality. To that end, the Chinese government has set the target of 5 GW and 30 GW of installed capacity by 2015 and 2030, respectively.

China's direct policy for offshore wind is composed primarily of fixed FiTs. The National Energy Bureau and the State Oceanic Administration classifies offshore projects as inter-tidal (the area between low and high tide along the coast line and with water depth of less than 10 metres), near shore (maximum 10 km away from the coast line with water depth between 10 and 50 metres), and deep sea (further than 10 km away from the coast line and water depth in excess of 50 metres) (Korsnes 2014). A national-level FiT of RMB0.75($0.12)/kwH for inter-tidal projects and of RMB0.85($0.14)/kWh for near-shore projects was announced in June 2014 (GWEC 2014). These FiTs will cover projects that come online before 2017, after which time the level may be reviewed by the government.

Although China has the largest onshore wind installed capacity in the world (more than 167 GW as of end of 2016), the progress towards achieving the offshore targets has been sluggish. It missed its 2015 target when the total installed capacity reached around 1 GW against the target of 5 GW. An inadequate level of FiT subsidies has been cited for the slow progress in the offshore wind industry. However, in 2016, the Chinese offshore market started what is believed to be its long-awaited take-off when 592 MW of new capacity was installed and pushed the cumulative capacity to 1627 MW (making China the third largest offshore wind-producing nation, after the UK and Germany). Nevertheless, there is little clarity on what the FiTs will be for projects consented post 2017, and the government's renewed focus on grid integration may also curb the growth of renewable capacity as the emphasis changes.

5.3.2.2 Japan

Following the Fukushima nuclear disaster in 2011, the interest in alternative energy and specifically offshore wind has increased considerably in Japan. Unlike other island countries such as the UK, Japan is surrounded by deep waters. Consequently, this fact constrains the potential of fixed-bottom off-shore wind farms to just a handful of sites. Therefore, Japan must rely primarily on the less-proven and costlier floating offshore wind turbine foundation concepts, and its policy support schemes must address this requirement.

To underpin investment in offshore wind projects, the government has established a relatively high fixed-rate FiT. Offshore wind power generators are eligible to receive an FiT of ¥36/kWh ($0.30/kWh) for 20 years, and the expectation is that this will create sufficient economic incentive for investors to enter in the market. Although the total installed capacity is just 60 MW as of year-end 2016, it should also be noted that the country has made tangible progress with demonstrating new floating wind turbine foundation technology concepts, suggesting that its current policy design is pushing the market forward.

5.3.2.3   South Korea

The first offshore wind turbine in South Korea deployed in 2012 and as of the end of 2016, the total installed capacity reached 30 MW. The country has a target of installing 2.5 GW by 2019 (Tsai et al. 2016) but it is not clear how this target is going to be achieved given the current level of progress. Moreover, difficult seabed conditions in the Yellow Sea have also hampered development of major projects, dampening expectations.

Prior to 2012, South Korea had an FiT scheme which supported investment in various small-scale renewable resources. However, in 2012, the FiT was replaced with an RPS (IEA 2016). This direct policy mechanism seeks to incentivise investments in renewable energy generation which are underpinned by guaranteed demand for electricity from renewable sources. Under the RPS model of South Korea, the 13 largest power companies with installed capacity in excess of 500 MW are required to steadily increase the share of renewables in their total power generation from 2% in 2012 to 10% in 2024. The eligible power companies will receive Renewable Energy Certificates (RECs) as compensation, which have a market value, based on their level of renewable generation.

To meet their obligations, the companies that are included in the RPS scheme need either to invest in renewable energy or purchase RECs in the market. Failure to surrender a sufficient number of RECs leads to a penalty that is 50% above the annual average price of RECs. To ensure fair competition in the market, technologies are banded whereby a multiplier is applicable to RECs of each band. For offshore wind farms that are installed at a distance less than 5 km from the coast, the multiplier is 1.5 and for those beyond 5 km the multiplier is 2, mimicking the former UK RO banded technology framework.

5.3.2.4   Taiwan

Given the land constraint, offshore wind power is an important part of Taiwan's renewable energy policy. Although the government has set the target of building 600 MW of offshore wind power by 2020, 3 GW of offshore by 2025, and a total of 4 GW by 2030, it is still far behind its neighbours.

To support its target of 600 MW of installed offshore wind, Taiwan's government has taken a phased approach by developing several small demonstration projects. In October 2013, grants were awarded to two projects: the 108 MW Fuhai wind farm and the 128 MW Formosa I wind farm and the government set a special FiT rate at TWD 5.56/kWh ($0.17/kWh) (GWEC 2014). Formosa I, developed by Swancor, is located in the Strait of Taiwan and is comprised of two phases. In phase one, a wind farm with a total capacity of 8 MW was installed in October 2016, largely to test out technologies and offshore construction. In phase II, an additional 120 MW is planned for installation in 2019, subject to final investment decision.

5.3.3   *United States*

The offshore wind power industry in the United States is still at an early stage of development. In December 2016, the first commercial offshore wind farm came into operation in America. The Block Island wind farm, which was deployed off the Coast of Rhode Island, includes five wind turbines with a total capacity of 30 MW.

While the outlook for the US market is generally more positive following the commissioning of the Block Island wind farm, the primary hindrances to policy development have been, on one hand, a lack of consistent federal support for offshore wind industry and, on the other hand, staggered election cycles that change out state-level and federal legislators. The political cycle has resulted in an unpredictable policy environment, and consistent political support for offshore wind has proven elusive and ultimately unbankable for most investors.

Policy developments on the state level will be key to supporting growth of the nascent US offshore wind industry. RO schemes (known as Renewable Portfolio Standards) on the state level mandate that regulated or licensed electricity suppliers source a proportion of electricity sold to customers from renewable energy sources. Credits for renewable energy generation are typically sold on markets for 'green certificates', with wholesale market prices varying—based on demand for renewable power generation. This mechanism provides an FiP on the wholesale price for offshore wind developers. However, RO frameworks vary on a state-by-state level. Some states have implemented specific 'carve-outs' for offshore wind in their RPSs, while others have not. This has resulted in a disjointed state-by-state level development approach, rather than a coordinated regional development effort that could spread out upfront infrastructure investment costs and support regional infrastructure and technology hubs.

Nevertheless, the trend of implementing RPSs is increasing. To date, some 29 states have RPSs for power sellers (National Conference of State Legislatures 2015). This figure includes most of the states in the US Mid-Atlantic and New England, where the offshore wind energy resources are among the strongest. Moreover, in certain states such as Massachusetts and New Jersey, new energy technologies including offshore wind are now featured just as prominently in the RPS as targets for existing technologies.

Federal tax credits are also potentially available to offshore wind developers. To spur the development of renewable energy infrastructure in the United States, The American Recovery and Reinvestment Act of 2009 had permitted offshore wind facilities placed into service by 31 December 2012 to receive an investment tax credit (ITC) worth up to 30% of capital investment expenditures. Since then, the ITC has repeatedly gone back and forth between expiration and being extended for one-year increments until the end of 2015. In December 2015, congress approved a multiyear extension of the renewable energy production tax credit (PTC) and ITC in the 2016 Consolidation Appropriation Act (DOE 2016). In this way, wind energy PTCs and ITCs were extended through 2016 at 100% of their 2015 value after which time their values decline 20% annually to reach 40% of its 2015 value in 2019. All projects started before the end of the period will qualify for that level of credits. Nevertheless, the stop-and-go nature of this incentive, particularly given the staggered election cycles in the US congress, has undermined the efficacy of this policy mechanism, as it is widely seen as unreliable.

5.4   EVALUATION AND DISCUSSION

As it can be seen from Fig. 5.3, although most of the industry growth has happened over the last 15 years, countries have had different degrees of success which place them at different stages of market development. Also, as is summarised in Table 5.3, the main adopted policy schemes around the world are variations of an FiP in the form of CfD, FiT, RPS, and ITC. The usage of a specific support policy, to a great extent, depends on the context, including institutions, market structure, technology maturity, and risk, amongst others.

The review of policy support mechanisms demonstrates that there are several options a government may choose from in terms of policy support models for offshore wind. While these may vary due to market structure

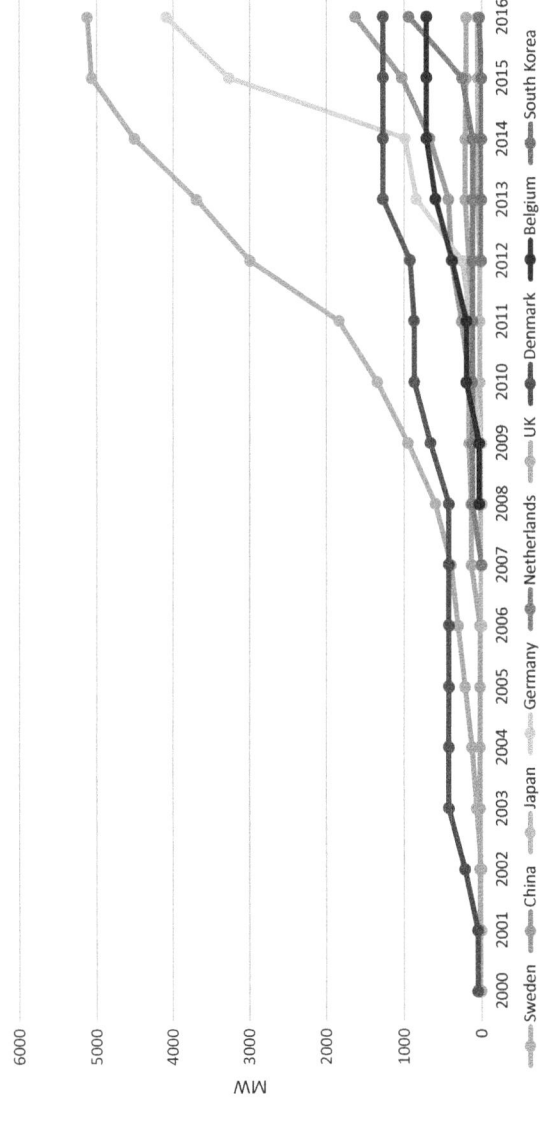

Fig. 5.3 The growth of offshore wind installed capacity in the main markets (Source: Authors)

Table 5.3 Summary of support policies in the main offshore wind markets

Country	Installed capacity (2016) (MW)	Policy mechanism features							Brief description of current policy support scheme
		Competitive tenders	Contract for difference	Feed in tariff	Feed in premium	RPS/Green Certificate	Investment tax credit		
United Kingdom	5156	◆	◆						The 'contract for difference' is an FiT/FiP model whereby the auction winners are entitled to the 'top up' payment. The top up payment is provided when the market price is below the strike price of CfD contract. When the market price is higher than the strike price, generators must return the difference
Germany	4019	◆	◆		◆				The FiP is calculated as the difference between market price and the maximum guaranteed price which traditionally was set administratively. Amendments to the German Renewable Energy Act (EEG) came into effect in 2017 and stated that offshore wind shall be subject to auctions and can no longer be remunerated through administrative prices
China	1627			◆					A national-level feed-in tariff of RMB 0.75 per kWh ($0.12) for inter-tidal projects and of RMB 0.85 per kWh ($0.14) for near-shore projects was announced in June 2014. These FiTs will cover projects that come online before 2017, after which time the level may be reviewed by the government
Denmark	1271	◆	◆		◆				An FiP that is the difference between guaranteed price and market price. The premium varies according to the market price and the guaranteed maximum price, which is determined through a reverse subsidy auction

(continued)

Table 5.3 (continued)

Country	Installed capacity (2016) (MW)	Competitive tenders	Contract for difference	Feed in tariff	Feed in premium	RPS/Green Certificate	Investment tax credit	Brief description of current policy support scheme
The Netherlands	1118	♦	♦		♦			A tender-based FiP that awards subsidy and consent to build and operate an offshore wind farm to the lowest cost developer for each zone as specified by the government. Offshore wind generators sell their electricity in the market and receive a premium as the difference between the maximum guaranteed price and the (corrected) market price. The maximum guaranteed price is determined through tender and is capped in order to ensure cost reduction
Belgium	712				♦	♦		The support scheme is based on a green certificate which is bought by the grid operator, Elia. The value of certificate is the difference between guaranteed set price (determined based on the average LCOE of offshore projects) and corrected market price which, in practice, makes the Belgian support model a form of FiP. The maximum price is currently set at €138/MWh and can rise to €150/MWh when the cost of grid connection is paid by wind farm
Japan	60	♦		♦				Offshore wind power generators receive a feed in tariff of JPY 36/kWh ($0.30/kWh) for 20 years, and the expectation is that this creates sufficient economic incentive for investors to enter in the market. Moreover, the offshore tariff is now 1.6 times higher than the onshore tariff, in order to improve investment in the sector

(continued)

Table 5.3 (continued)

Country	Installed capacity (2016) (MW)	Policy mechanism features						Brief description of current policy support scheme
		Competitive tenders	Contract for difference	Feed in tariff	Feed in premium	RPS/Green Certificate	Investment tax credit	
Taiwan	4			◆				Taiwan's government takes a phased approach by developing several small demonstration projects. In October 2013, grants were awarded to two projects: the 108 MW Fuhai wind farm and the 128 MW Formosa I wind farm and set a special FiT rate at TWD 5.56/kWh ($0.17/kWh)
South Korea	30					◆		In 2012, Korea replaced an FiT with a renewable portfolio standard. Under the RPS model, 13 largest power companies with installed capacity in excess of 500 MW are required to steadily increase the share of renewable in their total power generation from 2% in 2012 to 10% in 2024. The eligible power companies will receive Renewable Energy Certificates (RECs) as compensation, which have a market value
United States	30					◆	◆	In the United States, Renewable Obligation (RO) schemes on the state level mandate that regulated or licensed electricity suppliers source a proportion of electricity sold to customers from renewable energy sources. Credits for renewable energy generation are typically sold on markets for 'green certificates', which generate a market value. Some 29 states currently have RPS. There is no federal mandate

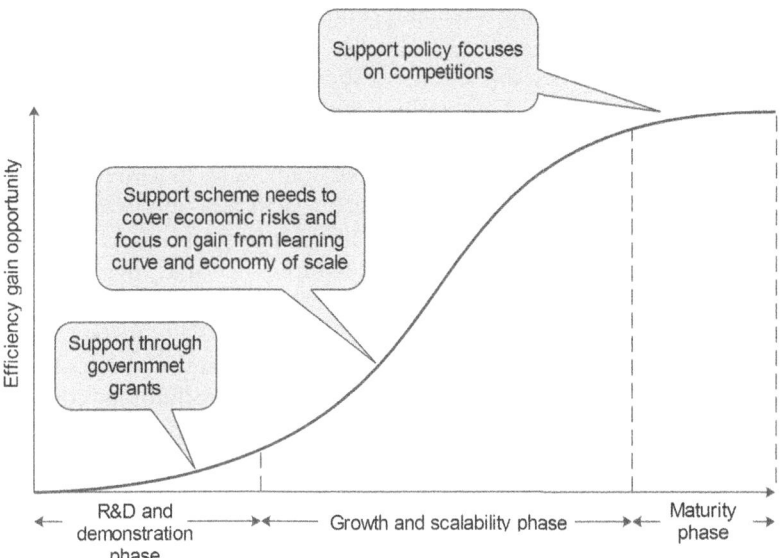

Fig. 5.4 The phases of renewable technology life cycles (Source: Authors)

(i.e., liberalised vs. centralised electricity market), the degree of technological maturity will also influence the type of policy mechanism that will be most effective.

The life cycle of offshore wind technology, as any other immature renewable generation, can be presented in three phases (see Fig. 5.4). The important policy consideration is that support scheme need to be suitable for the stage at which the technology is.

The first stage is research and development (R&D) and demonstration, which focus on showcasing the feasibility and performance of generation technology. As this phase is risky for private investors, governments' support in the form of grants for R&D and demonstration-scale projects is crucial. An example of this stage can be found in the nascent offshore industry of the United States, where the Department of Energy has awarded grants to offshore wind energy advanced technology demonstration projects.

The second phase is growth and scalability. During this stage, the technology enters the commercial stage; however, it is still costly without government support. The support policies at this stage often focus on lowering economic risks, promoting scalability, enhancing gain from the learning curve and economy of scale. The history of offshore wind power (and perhaps the wider renewable industry) shows that the most suitable support policy at this stage of technology is one that provides transparency, predictability, and security, in order to stimulate rapid and large-scale deployment of renewable technologies and contributes to lowering investment risk and financing costs. These are specifically important for less mature technologies such as that of the offshore industry.

The third stage is when technology matures such that the size of the industry is sufficiently large and there are many potential investors; however, it is not fully competitive at the prevailing electricity market prices. This is where the support policy should seek to exploit the power of competition to reduce the costs and, ultimately, make the technology competitive. Although the design of auction and the specifications of the adopted scheme may vary, overall, competitive policies are very effective in pushing down prices.

5.4.1   FiT and FiP Evaluation and Analysis

Among the current offshore wind support schemes, FiT and variations of FiP are the most popular models. The reason for popularity of FiT, as a policy to support offshore wind, is that it covers the economic risks and provides full insurance to investors. Furthermore, it is implementable under any market structure (monopoly, centralised, or liberalised). These specifications make it a suitable model for the growth phase of offshore wind technology. Currently, the nascent offshore markets of China, Taiwan, South Korea, and Japan (and France) all rely on FiTs to support their offshore industry.

Nonetheless, the disadvantage of the FiT approach, based on an administratively set price, is that setting the price at the efficient level is not straightforward given the presence of asymmetric information between offshore wind developers and regulators. The offshore wind power generators can be discouraged from investment when the tariff is set too low or alternatively may extract significant rent when the tariff is too high. This can even be the case when there is a regression mechanism in place. This is because it is difficult to set the price at levels that reflect the actual cost of development to the industry. This is why some countries combine FiT with the auction approach in order to avoid the challenges of setting prices administratively. Another problem of the FiT model is that it disconnects the production of a renewable facility from the market prices and thus can lead to over- or under-generation with consequences for grid stability.

In FiP models, despite the common feature of the subsidy payment as a 'top up' to the wholesale electricity market, there are significant differences in their designs. In some places, the subsidy is in the form of a certificate (Belgium, South Korea, and United States), whereas in other places, it is in the form of cash (UK, Denmark, and Germany). In some other markets, there is no guaranteed price (United States), whereas in most other markets, there is a maximum guaranteed price (all European countries). Along these lines, in some markets the maximum price is set administratively based on the LCOE of offshore projects (Belgium and Germany until recently), whereas in many other places the maximum prices are based on tenders (most other European countries).

The reason for the wide application of the FiP model, specifically in Europe, is that it fully integrates offshore wind power within the liberalised electricity markets and allows it to take advantage of electricity price movements. Furthermore, it makes renewable generators responsive to market prices and prevents them from producing when there is over-generation in the system. The other advantage of FiP is that if wholesale market prices increase because of carbon price or other reasons, the level of subsidies will decline when there is a maximum guaranteed payment. This is specifically relevant to European markets given that EU ETS is currently under reform, and the results of this reform may impact electricity market prices and pave the way for phasing out renewable subsidies.

However, the European FiP approach can impose additional cost of balancing service procurement on non-dispatchable resources, such as offshore wind, to respond to market price. Additionally, although the premium paid to offshore generators is floating in most European countries and these generators are entitled to a guaranteed tariff, there might still be profile risk with respect to the way that reference market price is specified.

The US support model of green certificates, which is a variation of FiP, exposes generators to even more risks compared with European models. This is because in the US model, generators need to deal with two sources of price volatility: the electricity wholesale market and the market for green certificates. However, in the United States, offshore wind developers benefit from ITCs and PTCs which can also be significant.

5.5   Conclusions

This chapter reviewed the support policy mechanisms that have been adopted in the main offshore wind markets around the world. The results of our investigation show that FiT and variations of feed-in premium are the two most widely adopted models to incentivise investment in offshore wind technology. Fixed FiT is a popular model in the Asian offshore markets (except South Korea), which are in their expansion phase. Most European countries, on the other hand, now use a form of FiP wherein the subsidies have been linked to the liberalised electricity markets. In addition, most European offshore wind markets have already moved or are considering a move towards competitive support policies.

The examination of offshore wind markets around the world shows that the design of an effective support scheme depends on the specifics of the market and the degree of technology maturity. There are three stages discernible for offshore wind technology life cycle: proof of concept and demonstration, scalability and growth, and, finally, maturity.

At the early stage, when technology is being tested to showcase its performance and feasibility, government grants that contribute towards capital costs are the most suitable policy for supporting the industry. When the technology enters the commercial phase, where the aim is to increase the share of technology in the generation mix and reduce the cost, the most effective support policy is a scheme that shields investors from market price, provides certainty of future revenue, and minimises off-take risks by offering a long-term revenue guarantee that can be accompanied with a Power Purchase Agreement (PPA). In mature markets, where size of the industry is sufficiently large, competitive policies can be employed in order to exploit the power of competition and incentivise cost reduction across the whole supply chain. The important policy consideration is that support schemes need to be compatible with the stage of technology during its three-part life cycle.

References

BEIS. (2016). *Draft Budget Notice for the Second CFD Allocation Round*. Department for Business, Energy and Industrial Strategy. https://www.gov.uk/government/uploads/system/uploads/attachment_data/file/566307/Draft_Budget_Notice_FINAL.pdf. Accessed 3 Mar 2017.

BMWi. (2015, February). *The Energy Transition – A Great Piece of Work. Offshore Wind Energy: An Overview of Activities in Germany.* Federal Ministry for Economic Affairs and Energy (BMWi). https://www.erneuerbare-energien. de/EE/Redaktion/DE/Downloads/offshore-wind-energy.pdf?__ blob=publicationFile&v=2

DECC. (2015). *Budget Revision Notice for CfD Allocation Round 1.* Department of Energy and Climate Change. https://www.gov.uk/government/uploads/ system/uploads/attachment_data/file/398665/150127_Budget_Revision_ Notice_for_CfD_Round_One.pdf. Accessed 3 Mar 2017.

DOE. (2016, September). *National Offshore Wind Strategy: Facilitating the Development of the Offshore Wind Industry in the United States.* US Department of Energy, DOE/GO-102016-4866. https://energy.gov/sites/prod/ files/2016/09/f33/National-Offshore-Wind-Strategy-report-09082016.pdf

GWEC. (2014). *Global Wind Annual Market Update.* Global Wind Energy Council. Retrieved from www.gwec.net/: www.gwec.net/wp-content/ uploads/2015/03/GWEC_Global_Wind_2014_Report_LR.pdf

IEA. (2016). *Renewable Portfolio Standard (RPS).* International Energy Agency. http://www.iea.org/policiesandmeasures/pams/korea/name-39025-en.php ?s=dHlwZT1yZSZzdGF0dXM9T2s. Date Accessed 9 Mar 2017.

Korsnes, M. (2014). *China's Offshore Wind Industry 2014: An Overview of Current Status and Development.* Norwegian School of Science and Technology. CenSES Report 1/2014. https://www.ntnu.no/documents/7414984/ 202064323/Offshore+Wind+in+China+2014.pdf/ b0167dd4-6d47-40cc-9096-b3139c1459ef

National Conference of State Legislatures. (2015). *State Renewable Portfolio Standards and Goals.* www.ncsl.org/research/energy/renewable-portfolio-standards.aspx. Accessed 19 Feb 2015.

NEA. (2015, January). *Offshore Wind Energy in the Netherlands: The Roadmap from 1,000 to 4,500 MW Offshore Wind Capacity.* Netherlands Enterprise Agency. https://www.rvo.nl/sites/default/files/2015/03/Offshore%20 wind%20energy%20in%20the%20Netherlands.pdf

OffshoreWIND.biz. (2016). *Vattenfall Wins Kriegers Flak Tender with EUR 49.9 per MWh Bid.* http://www.offshorewind.biz/2016/11/09/vattenfall-wins-kriegers-flak-tender-with-eur-49-9-per-mwh-bid/ Posted on November 9, 2016. Accessed 8 Mar 2017.

TKI Wind OP ZEE. (2015). *Subsidy Schemes and Tax Regimes.* TKI Wind op Zee (Top Consortium for Knowledge and Innovation Offshore Wind). Retrieved from http://tki-windopzee.eu/files/2015-09/20150401-rap-subsidy.and. tax.policies-pwc-f.pdf

Tsai, Y., Huang, Y., & Yang, J. (2016). Strategies for the Development of Offshore Wind Technology for Far-East Countries—A point of View from Patent Analysis. *Renewable and Sustainable Energy Reviews, 60,* 182–194.

UK Government. (2015, February 26). *Contracts for Difference Allocation Round One Outcome.* Retrieved from www.gov.uk: https://www.gov.uk/government/uploads/system/uploads/attachment_data/file/407059/Contracts_for_Difference_-_Auction_Results_-_Official_Statistics.pdf

Watson Farley & Williams. (2016, July). *Energy Briefing: The New German Offshore Wind Act,* Publication code number: 58565672v4© Watson Farley & Williams. http://www.wfw.com/wp-content/uploads/2016/07/WFW-Briefing-Germany-WindSeeG-2017_EN-July-2016.pdf

CHAPTER 6

Cost Reductions and Innovation in the Offshore Wind Industry

Abstract Offshore wind is a relatively immature industry which is heavily reliant on government subsidies to grow. In order to survive in an environment where the cost of alternative renewable energies, such as onshore wind and solar, have been rapidly falling, the offshore wind industry must significantly reduce its costs in the coming years. This chapter discusses the issue of cost reduction and innovation in the industry and analyses the opportunities for cost cutting through improving technology, supply chain management, finance, and government support policies.

Keywords Offshore wind cost reduction • Innovation • Finance • Supply chain management • Support policies

6.1 Introduction

The offshore wind industry was born in the mature and established energy markets of Europe, where fully developed cheap generation technologies were readily available. Up to now, the growth of this industry has therefore been heavily reliant on government subsidies, which are considered to be costly to the public budget and can be susceptible to political intervention. Therefore, in order to have a future and survive in a competitive environment, cost reduction has become a top priority for the offshore wind industry and policy makers. There are at least four areas that can

© The Author(s) 2017   91
R. Poudineh et al., *Economics of Offshore Wind Power*,
https://doi.org/10.1007/978-3-319-66420-0_6

contribute to the objective of cost cutbacks in this industry. These are technology, supply chain management, finance, and support policies (see Fig. 6.1). An effective strategy to lower the cost of the offshore wind industry entails focusing on all available options.

An important domain of cost reduction is the technology segment of offshore wind, including the turbine, foundation, grid interconnection, installation, and operation and maintenance (O&M). Moreover, the offshore wind supply chain has potential for cost cutbacks. This is reflected in the low level of integration among supply chain actors, lack of standardisation, lack of transparency of demand, inefficient order and inventory process, and lack of collaboration across supply chain (Stentoft et al. 2016).

Another major area of potential cost trimming is finance. The cost of capital constitutes the lion's share of offshore wind project costs. The presence of risk and uncertainties can adversely impact the cost of capital and lower the appetite of lenders to become involved in this industry. Finally, government support policies play a crucial role in lowering the costs in the industry.

Innovation in any of the aforementioned four domains of the offshore wind industry, whether it is in the form of a new product, process, technology, or business model and finance method, can lead to cost saving through three channels: reducing operation and maintenance expenditure (OPEX), cutting capital expenditure (CAPEX), and/or increase in annual energy production (AEP).

This chapter focuses on the issue of cost reduction in the offshore wind industry with an emphasis on technology, supply chain, finance, and government support policies. The next section discusses both historical and potential future innovations in the technology segment of the offshore wind industry, whereas Sect. 6.3 presents the opportunities for cost trimming through supply chain management. Section 6.4 investigates the role of finance in lowering costs before Sect. 6.5 analyses the contribution of government support policies in making offshore wind energy competitive. Finally, Sect. 6.6 concludes.

6.2   Technology

6.2.1   Turbine

Wind turbine technology has improved considerably since the inception of the offshore wind industry. Currently, the power rating of new turbines is

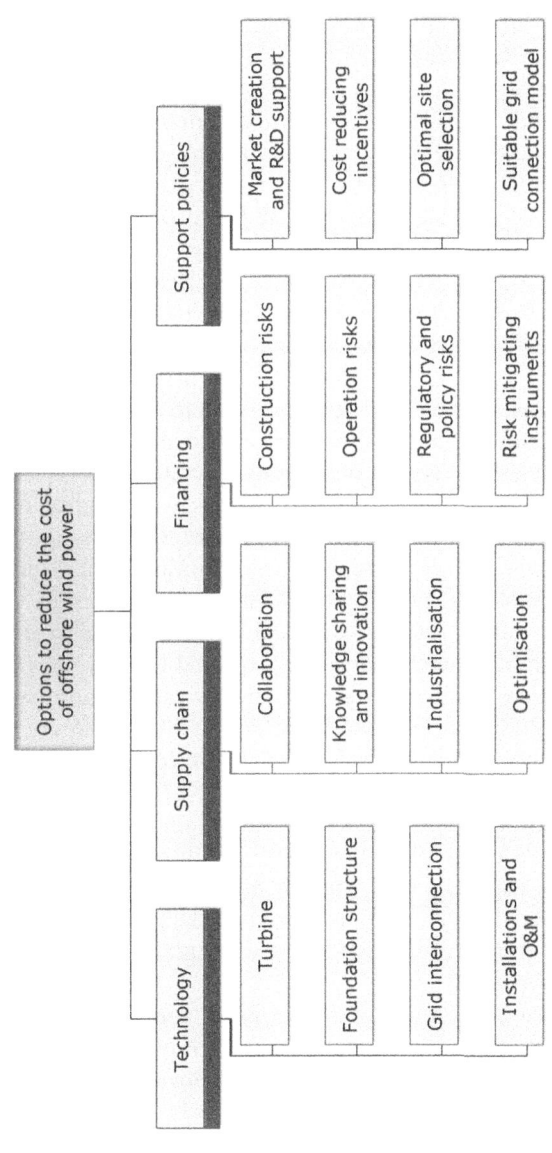

Fig. 6.1 Options to reduce the cost of offshore wind power (Source: Authors)

significantly higher than that of the earlier models. The growth in the size of rotor, improvements in design and manufacturing of blades and drive-train technology, along with introduction of structural innovations such as the use of carbon fibre have all enabled greater energy capture without increasing the cost per unit of power ($/MW) (IRENA 2017). Better manufacturing and the use of sophisticated design tools have also enhanced the torque-level tolerance of turbines and improved reliability of components, resulting in a decline in O&M costs and extended operational lifetimes. The last decade has also witnessed improved aerodynamics and controls in turbines which have reduced loading and increased the energy capture capacity. However, the cost effects of these different innovations are not similar; the largest cost saving as a result of innovation in turbine technology is related to the improved power ratings and increased rotor sizes.

Innovation in turbine technology is expected to continue in the coming decades. There are currently ongoing works to develop and test a new generation of drivetrains, superconducting generators as a replacement to traditional copper winding generators, blades with higher tip speeds, downwind turbines, new approaches to hub assembly components, and advanced electricity take-off systems. These innovations not only improve reliability and lower O&M costs, but also allow for the design and manufacture of higher rating turbines with significantly greater cost efficiency resulting from higher energy yields. It is expected that turbines with rated capacities of approximately 10 MW and 15 MW can be commercialised in the 2020s and 2030s, respectively. The recent trend also confirms that reaching such capacities is not far from reality. In May 2017, Danish company DONG Energy finished installing a wind farm consisting of 8 MW turbines in Liverpool Bay of the United Kingdom. There are also efforts to reduce material usage of turbines through introduction of two-blade turbines which lowers the cost without a proportional reduction in the energy capture capacity. The two main sources of efficiency gain in turbine technology over the next two decades are expected to be larger rotors and life extension beyond 30 years.

Since the turn of century, innovations in turbine technology have reduced the levelised cost of energy (LCOE) of offshore wind power by around 20%, the lion's share of which pertains to OPEX rather than CAPEX (see Fig. 6.2). This is also expected to be the case over the next 15 years, albeit at a slower rate. The improvement in AEP has contributed to reduction of LCOE by 18% in the last 15 years. However, in the com-

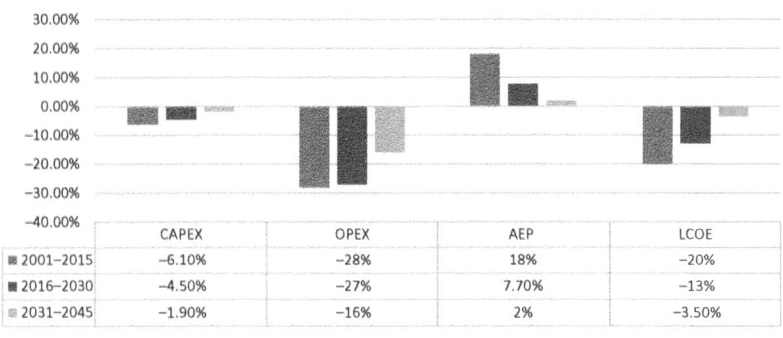

	CAPEX	OPEX	AEP	LCOE
■ 2001–2015	−6.10%	−28%	18%	−20%
■ 2016–2030	−4.50%	−27%	7.70%	−13%
▨ 2031–2045	−1.90%	−16%	2%	−3.50%

■ 2001–2015   ■ 2016–2030   ▨ 2031–2045

Fig. 6.2 The components of historical and future expected efficiency gain from turbine (Source of data: IRENA (2017))

ing decades, this opportunity is expected to be exhausted once the size of turbines reaches the optimum point. The reduction of CAPEX, however, has been challenging because turbines are material intensive and the cost increases, albeit at a slower rate, as the size and rating of turbine grow.

6.2.2   Foundation Structure

The types of foundations that can be used for offshore wind turbines depend on various factors that include seabed condition, water depth, turbine mass, rotor speed, loading of turbine, developer expertise, and capabilities of the supply chain. To date, the main types of foundations that have been used are monopiles, jacket or other space frame foundations, and gravity-based foundations, all of which are fixed to the bottom of the sea. Among these models, monopiles have been, by far, the most widely adopted foundation type.

The design and manufacturing of foundations has changed over time. In the early years of the industry, most foundations were designed by offshore oil and gas consultants and were manufactured in oil and gas fabrication yards and shipyards. The standards of foundation design were also primarily the same as offshore oil and gas structures. These standards were developed to design structures which had significantly higher asset values, significantly higher environmental risks in the event of failure, and, most importantly, were manned with personnel more often than not. As such, they included a much higher life safety risk. These standards, however, are considered too conservative and, therefore, costly for the offshore wind industry both in terms of design and fabrication (IRENA 2017). Some of these issues are being effectively mitigated. Over time, dedicated manufacturing yards were established to enable serial production of foundation components and reduce the cost. Also, the data obtained from performance observation of installed foundations has allowed developing design standards that are more relevant to the offshore wind industry. This is also the case with respect to the design of purpose-built transportation and installation vessels. The new design standards lowered the amount of steel material consumption and an improved economy of scale, and thus reduced the costs. As seen from Fig. 6.3, since 2000, these innovations have mainly reduced capital costs as opposed to operating expenditures.

In the coming decades, it is expected that design standards will improve further through, for example, saving on steel content and installation time, serial fabrication of foundations, introduction of self-installing (crane-less) foundations, and taking a holistic approach to design concepts (combining the best feature of various designs). Figure 6.3 shows that the highest cost-saving effect on LCOE from foundation designs is expected to happen in the next two decades, despite the fact that in the past 15 years this area has not witnessed much efficiency gains. Also, innovations in foundation design do not have any improvement effect on AEP.

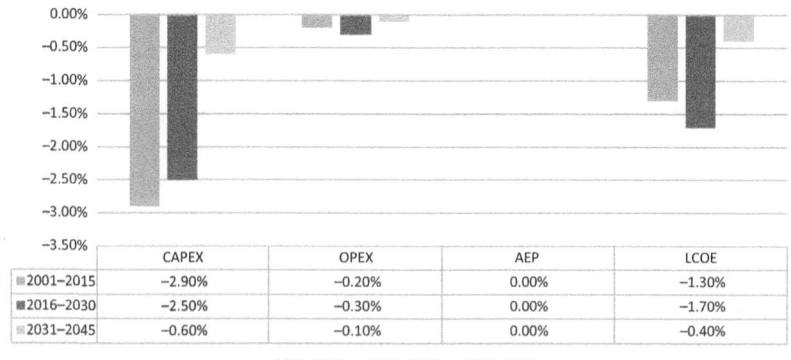

	CAPEX	OPEX	AEP	LCOE
2001–2015	−2.90%	−0.20%	0.00%	−1.30%
2016–2030	−2.50%	−0.30%	0.00%	−1.70%
2031–2045	−0.60%	−0.10%	0.00%	−0.40%

■ 2001–2015 ■ 2016–2030 ■ 2031–2045

Fig. 6.3 The components of historical and future expected efficiency gain from foundation structure (Source of data: IRENA (2017))

6.2.3   *Electrical and Grid Equipment*

The grid interconnection infrastructure includes the submarine cables (array cables) that collect power from the wind farm, the offshore substation (collection point), and the export cable to the shore. Over the years, advancements in the monitoring of electrical systems have led to more effective control of active and reactive power and active control of power factor compensation. These, in turn, have improved reliability and lowered the cost of grid interconnection. Moreover, the usage of high-voltage export cables and efficient conductor materials has resulted in a reduction of energy losses and lower costs. Since 2001, these innovations have reduced the LCOE by 2.8% (see Fig. 6.4).

There are several areas that technology of grid interconnection may further improve in the future (IRENA 2017). In relation to cables, choosing the optimum core size, insulation thickness, and protection equipment will lead to weight reduction and a consequent efficiency gain. There is also the possibility of raising the operating voltage of array cables and specifically using high-voltage direct current (HVDC) cables for power export over long distances. This increases the capacity, reduces energy losses, and provides the opportunity to take advantage of windy sites far from the shore. With regard to substations, efforts are being made to use modular structures and floating designs that negate or reduce the need for heavy-lift installation vessels. In addition, innovations that enable the installation of electrical equipment directly in selected turbine towers

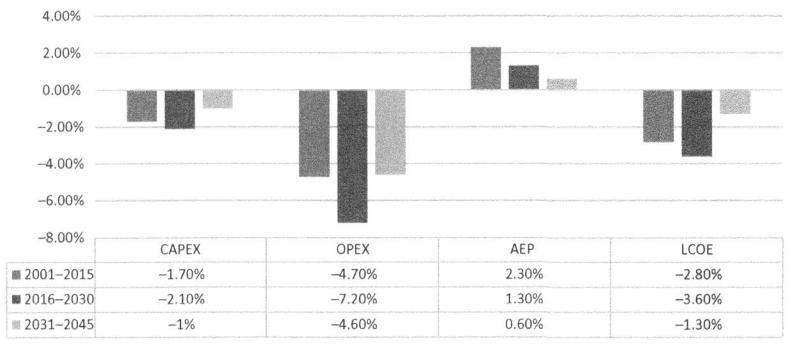

	CAPEX	OPEX	AEP	LCOE
■ 2001–2015	−1.70%	−4.70%	2.30%	−2.80%
■ 2016–2030	−2.10%	−7.20%	1.30%	−3.60%
▨ 2031–2045	−1%	−4.60%	0.60%	−1.30%

■ 2001–2015   ■ 2016–2030   ▨ 2031–2045

Fig. 6.4 The components of historical and future expected efficiency gain from grid interconnection (Source of data: IRENA (2017))

rather than on an independent offshore substation are under way. This particular innovation is expected to constitute the greatest source of cost saving for grid interconnection over the next two decades.

The regulatory model of grid connection is also important because it governs the distribution of costs between project developers and grid operators or third parties. Currently, there are three main regulatory approaches for the treatment of offshore grid connection costs. The first one is a generator model, which requires wind project developers to bear the entire cost of grid connection. In such a model, wind farm developers have a high incentive to implement a cost-efficient connection because high cost or low availability directly affects their profits from the wind farm (Green and Vasilakos 2011). At the same time, a generator model significantly increases the project developers' costs. This model is currently being practised in some countries, such as Sweden. The second approach is based on the idea that the transmission system operator (TSO) is responsible for extending the grid in order to reach the wind farm. This model is the dominant method for onshore grid connections and several countries, such as Germany and Denmark, have extended it to their offshore projects.

The third approach is the UK model, in which a tender is run in order to appoint a third party as the offshore transmission owner (OFTO). The OFTO would then be responsible for building, owning, and operating the connection asset between the wind farm and the mainland, bound by a set of standards and codes applicable to the industry. This model, which launched in 2009, entitles OFTO licence holders to 20 years of revenue stream, subject to a satisfactory performance, indexed to the retail price index (RPI) in the United Kingdom. The OFTOs' revenue, which comes from the National Electricity Transmission System Operator (NETSO), is independent of wind farm performance, as the transmission asset owner is only required to ensure its availability, irrespective of actual power transmitted. As the United Kingdom was previously working under a generator model, the OFTO regime applies both to the transmission assets acquired from the wind farm developers and to the transmission assets newly built by OFTOs. The first licence for an OFTO was granted in 2011, and by March 2014 there were nine operating OFTOs (Ofgem 2014).

The key advantage of the OFTO approach is that it allows new entrants to enter the market and thus can deliver cheaper and timelier offshore grid connections through the enabling of competition. Additionally, the OFTO model focuses on a generator's needs and provides flexibility for future offshore generation requirements. Also, this approach is subject to light-handed regulation and is protected against generator failure and credit risk. However, despite these appealing characteristics, Meeus (2015) argues that, in the case of the OFTO and generator models, several entities are involved in the process of design and development of the offshore grid connection. Instead, the TSO model involves fewer parties, and the related institutions already exist and are mature. Nonetheless, due to the absence of competition in the TSO model, it is not suitable for large-scale offshore wind integration at some distance from the coast. Consequently, a trade-off arises between opting for a more competitive but complex model (e.g., the OFTO approach) and a more straightforward but probably less efficient regulatory model for offshore wind farm grid connection (e.g., the TSO model).

There is also a huge potential for regional cooperation in building the offshore interconnections. In Europe, for example, the issue of transmission infrastructures in the North and Baltic seas is high on the agenda of policy makers, mainly because of the anticipated large-scale offshore wind farm projects. In this respect, the North Sea Countries Offshore Grid Initiative (NSCOGI) was formed in 2010 (NSCOGI 2014) to facilitate cooperation among relevant EU countries in order to build an integrated interconnection for offshore wind farms as well as other renewables. Such a cooperation can provide significant cost efficiency for grid connection. The key challenge, however, is to overcome the regulatory, planning, and economic obstacles of connecting offshore wind farms to the network.

6.2.4   Installation and O&M

Installation activities along with O&M are other areas with the potential for cost reduction. The offshore installation activities include foundations, turbines, and grid interconnections. In the past, these activities have been carried out under separate contracts. In recent years, it has become commonplace to undertake all engineering, procurement, construction, and installation (EPCI) in one contract. The lessons learned of the past two decades have led to some innovative cost reduction installation strategies. These include, for example, the use of two vessels (expensive and cheap) for piling and transition piece installations, introduction of pre-piling for jacket foundations, introduction of specialised vessels for cable laying, and introduction of flexible sea fastening as opposed to previous method of component-specific sea fastening. The use of optimum vessels during each

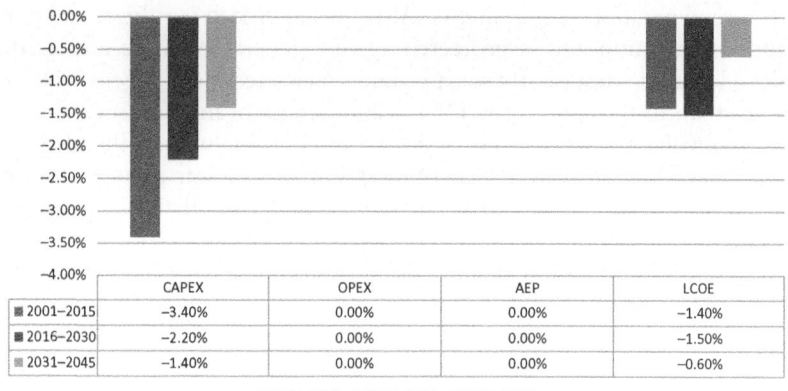

	CAPEX	OPEX	AEP	LCOE
■ 2001–2015	–3.40%	0.00%	0.00%	–1.40%
■ 2016–2030	–2.20%	0.00%	0.00%	–1.50%
▤ 2031–2045	–1.40%	0.00%	0.00%	–0.60%

■ 2001–2015   ■ 2016–2030   ▤ 2031–2045

Fig. 6.5 The components of historical and future expected efficiency gain from installation (Source of data: IRENA (2017))

installation phase has been the main source of cost reductions for installation activities (IRENA 2017). As seen from Fig. 6.5, installation innovation reduces LCOE only through saving on capital expenditures.

Potential future innovations in installation processes include the development of high-wind-speed blade installation techniques, further optimisation of vessels according to installation phase, speeding up the process of connecting array cables to the wind farm, and the development of integrated turbine installation to eliminate some of the intermediary steps (foundation is installed initially and then turbine is installed in one step).

O&M constitutes all activities that start from completion of the project installation and commissioning to the facility decommissioning at the end of the useful lifetime. This includes the management of contracts and operations, onshore facilities, offshore logistics, wind turbine, and balance-of-plant maintenance (both planned and unplanned) (IRENA 2017). These activities can be fully or partially delegated to others through contract or can be carried out by the wind farm owner. The way in which O&M is executed has evolved considerably in the last two decades. The advances in condition monitoring and prognostics have resulted in better turbine availability and lower operating costs. Also, the use of larger personnel transfer vessels, introduction of special service operation vessels, along with the use of helicopters, have widened the weather condition window for wind farm accessibility. This has resulted in lower operating

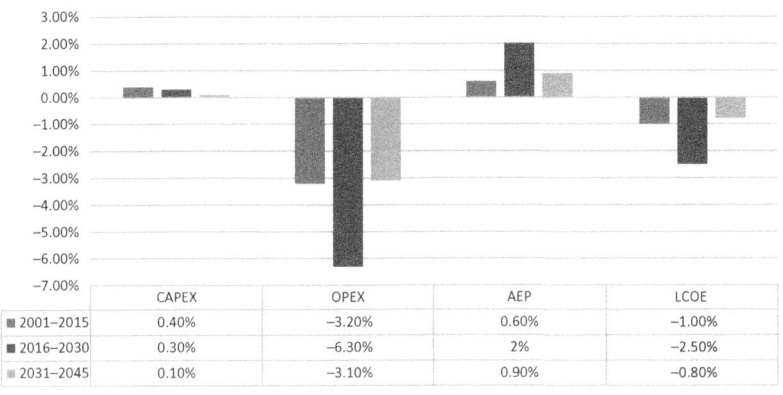

	CAPEX	OPEX	AEP	LCOE
■ 2001–2015	0.40%	−3.20%	0.60%	−1.00%
■ 2016–2030	0.30%	−6.30%	2%	−2.50%
■ 2031–2045	0.10%	−3.10%	0.90%	−0.80%

■ 2001–2015   ■ 2016–2030   ■ 2031–2045

Fig. 6.6 The components of historical and future expected efficiency gain from O&M (Source of data: IRENA (2017))

costs and a higher turbine availability. During the last 15 years, the LCOE has fallen by around 1% as a result of more efficient O&M. But, it is likely to decrease further in the coming decades (see Fig. 6.6).

In the future, it is expected that advancement in weather forecasting, the introduction of condition-based as opposed to time-based maintenance, the introduction of remote and automated maintenance, and the use of drones for aerial inspection of turbines can all contribute towards lowering the LCOE. As can be seen from Fig. 6.6, more reliance on technology in the future can lead to a slight increase in capital costs but the net effect is a reduced LCOE due to a significant decline in OPEX.

6.3   Supply Chain Management

The supply chain is an important source of cost reduction, and thus it has always been a target for innovation incentives. Along with competition among supply chain members, collaboration plays a key role in the development of more efficient products and processes.

The physical supply chain of the offshore wind industry starts with the turbine (Stentoft et al. 2016). The turbine includes three main components: rotor, nacelle, and tower. The turbine requires a support foundation which, depending on the specifics of the seabed and wind farm, can use different design concepts. Then, there is the grid interconnection, which is provided through subsea cables and an offshore substation that exports high-voltage electricity to the shore. The installation of the foundation, turbine, and grid interconnection is carried out by manned vessels. This constitutes the major supply chain elements up until commissioning, at which point the O&M supply chain takes over. Towards the end of the life cycle of a wind farm, the repowering or decommissioning supply chain becomes integral.

In reality, however, the offshore wind supply chain includes several intersecting supply chains that are made up of various components and subsystems which are produced by large-scale manufacturers as well as small manufacturers. Overall, there are three ways that supply chain management can reduce the cost of offshore wind farms: innovation, industrialisation, and supplier partnering (Stentoft et al. 2016). Cost-reducing innovations in the supply chain include both product and process innovations.

Innovation is knowledge based. However, this knowledge is dispersed across many firms and stakeholders across the industry. There are several ways that the offshore wind industry can leverage the unique competency of the supply network. First, value chain integration can help firms in the offshore industry to develop a sourcing strategy, set clear goals, and engage in a high degree of information exchange with suppliers to increase end-to-end visibility. Second, knowledge sharing and knowledge management across the supply chain can help the offshore wind industry to increase innovation in new technologies and new products that can reduce the cost of energy. Third, value chain mapping can be employed to eliminate waste and non-value-added activities in the supply chain operation. Waste in the supply chain can be due to inefficient use of time, movement, and material flow as well as inefficient decision making. Fourth, supply network architecture can provide great potential for cost reduction. It includes supplier and source selection, the degree of autonomy given to the first-tier suppliers, contract management, and degree of personal and professional ties among members of the supply network. Fifth, supply chain analytics can be used by the offshore wind industry to extract customer and supply market intelligence both from quantitative and qualitative data in order to facilitate cost-reducing innovations.

Another important area of cost reduction through supply chain management is industrialisation which, in this context, means the ability of a firm or an industry to efficiently manage its internal and external resources. This, in turn, depends on the stage at which the industry is in the maturity curve. The most common practices in mature industries from which the offshore industry can learn are cost modelling, supplier development initiatives, and supply chain process optimisation. Cost modelling can help firms in the offshore industry make efficient sourcing decisions by developing an independent understanding of cost structures of needed components separate from suppliers' quotes. The supplier development initiative is a process that includes training personnel in cost modelling and statistical process control, quality management, and continuous improvement practices. On the other hand, the supply chain process optimisation implies executing processes with maximum efficiency by, for example, increasing responsiveness to partners, enhancing the accuracy of process outcomes, and standardisation of the process.

Finally, partnering is perceived to be an effective approach to reduce the costs of the supply chain. In mature industries, it includes joint development of technology, sharing of products and technology strategies with suppliers, flexible contracts, and a high degree of IT integration. In some cases, deep collaboration with suppliers can lead to specific investment and commitment to availing specialised skills and human capital by the supplier firm for the buyer.

6.4   Finance

The cost of capital constitutes a significant share of the LCOE of offshore wind farms. The capital spent at the inception of a project needs to be paid back over time and, given that money has a time value associated with it, the ultimate implication is that this capital is more expensive than if it were tapped more over the course of the operating life of the asset (Krupa and Poudineh 2017). Furthermore, any uncertainty in investment return and/ or payback period can increase the cost of capital. Although the cost of capital is unavoidable, it can be minimised by lowering the project risks and improving its financeability. For example, if the project were built overnight and risk free, it would cost much less.

For the purpose of financing, large offshore wind projects are usually developed through a standalone project company, which is owned by project investors and has its own revenue and balance sheet. Generally, there are two sources of finance for offshore wind projects: corporate finance and project finance. In corporate finance, the project is financed through the balance sheet of the parent company and the finance is based on the risk profile of the main company as a whole, and not the specific project itself. This financing method, which is traditionally preferred by large utility companies with strong balance sheets, often results in lower risks and a consequent lower cost of capital. In the project finance approach, however, the main sources of capital are lenders (i.e., banks) and the cash flow of the project determines the key financial parameters. Although in this approach the project assets, rights, and interests are held as secondary security or collateral, lenders have no recourse to the assets of sponsor companies. In practice, the financing structure of offshore wind is slightly more complex. For example, sponsor companies may create a joint venture to enhance the financeability or, for taxation and other legal purposes, form separate special-purpose vehicles (SPVs) or holding companies through which they raise project finance. Figure 6.7 presents

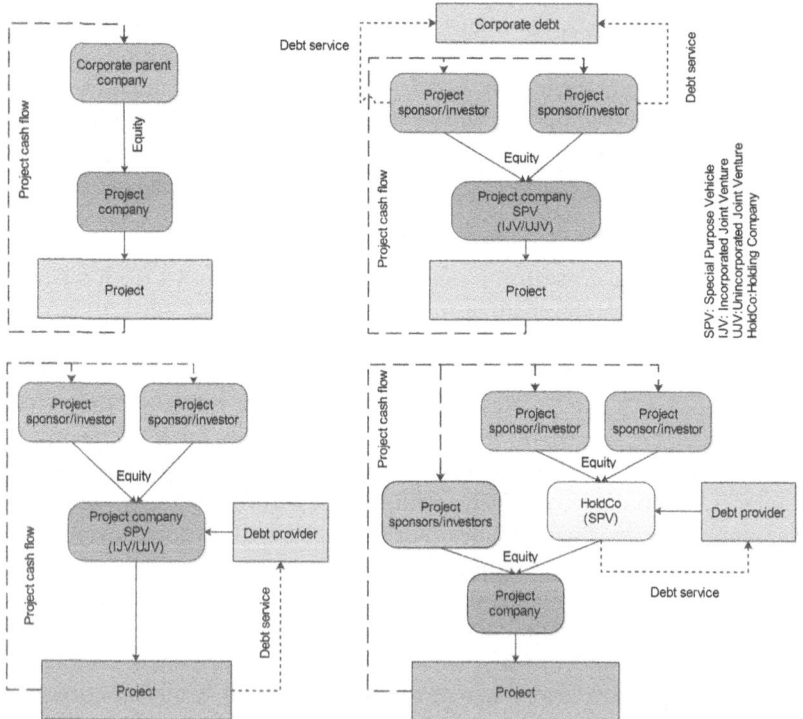

Fig. 6.7 Typical financing structure of offshore wind (Source: Authors adapted from FTI consulting (2015))

typical finance structures of offshore wind farm projects. The finance model of the project not only has an impact on the cost of capital, but also the way that project needs to deal with contractors given the bank requirements.

In order to reduce the cost of financing for offshore wind, key risks and uncertainties need to be identified and appropriate risk mitigation strategies need to be adopted. Overall, there are three types of risks that offshore wind farms are exposed to: construction risk, operation risk, and regulatory risk.

The construction phase of an offshore wind farm is a complex process and finance providers perceive grid interconnect and reliability of suppliers as the most important sources of risk during this period (EWA 2013). The possibility of delays in grid construction and connection with consequent implications for project overrun is a major concern. This is specifically exacerbated when the party in charge of grid construction is different from that of grid connection. The grid issue specifically has impacted the willingness of debt providers to accept construction-phase risks. In the United Kingdom, by allowing the wind farm to control grid delivery in the transitional period (before operation), decoupling grid revenue from generator output, and categorising the grid as a separate asset class, the grid construction risks have been reduced significantly. This is because the de-risked offshore grid itself has become an attractive investment which can attract a large pool of finance.

The suppliers' risks include credit risk, contracting risks, and installation and logistics risks (EWA 2013). The credit and financial situation of a supplier determine its ability to fulfil its commitment and provide a guarantee when the equipment or service does not function the way it is intended. Multi-contracting, with no dominant counter party, is another source of risk which can specifically lead to intrusive due diligence from debt providers to reduce interface risks. Nonetheless, some argue that a well-structured multi-contract can be less risky than a full EPCI contract with a general contractor that may not undertake sufficient due diligence on its subcontractors. Overall, some measures can be taken to reduce these risks, such as strong contractual provisions, interaction with reputable contractors, and the use of teams with expertise in interface management.

Installation and logistic risks are important in the construction phase due to the challenge of operating in the harsh marine environment. The assessment of lessons learned from industry experience and implementation of best practices are important ways of dealing with these risks. However, appropriate policies need to be in place in order to ensure knowledge sharing. Furthermore, the selection of contractors with knowledge of local construction along with setting aside sufficient capital for unforeseen circumstances can lower installation and logistic risks.

Operation risks include technology and reliability risks (EWA 2013). The use of new and unproven technologies for blades, gearboxes, and other parts of a wind farm poses a technology risk which may cause the finance providers to take a conservative approach. At the same time, the possibility of component failure constitutes another source of risk. However, these risks, to some degree, can be addressed through guarantees and liquidated damage clauses in purchase contracts with equipment manufacturers. Manufacturers can also provide more information to lenders on the reliability of equipment or participate in the project as a minority shareholder.

Finally, regulatory and policy stability is perhaps one of the largest risks facing the offshore wind power industry. This includes the retroactive adjustment to renewable support schemes or changes to requirements on balancing obligations for renewables, priority dispatch, and other network codes. The presence of regulatory uncertainties has an adverse effect on project risk and lowers the appetite of lenders to provide capital to offshore wind projects. These all highlight the importance of governments honouring their commitment to the support of the industry. The lack of commitment has a serious impact on investors' and lenders' confidence. The mechanism for policy change needs to be well designed and actual adjustment should be implemented only after consultation with stakeholders. New policy must not make the existing beneficiaries of renewable support schemes worse off or better off.

6.5   Support Policies

Government policies including subsidies, taxes, site selection, incentives for efficiency, and innovation and procurement methods, each play a critical role in reducing the cost of offshore wind farms. By creating an attractive market in which there is a viable business model and a clear pipeline of projects, policy can reduce the cost of produced energy through its impacts on technology, finance, and the supply chain. An attractive market can be created through offering long-term guaranteed revenues that provide a competitive return to project investors. This is specifically important given the interdependency between the market size and costs. The efficiency gains from the learning curve, economies of scale, and competition can be realised only if the industry, as a whole, invests in the new products, technologies, and techniques. This, in turn, is contingent upon the existence of a supportive policy framework which ensures a sustainable offshore wind market with a sustained growth.

The introduction of competitive tenders in the mature markets along with setting a maximum bid and targets for cost reduction is a measure that has proven to be effective in materialising cost efficiency. The experience with tenders in the mature offshore wind markets of Europe shows that a well-designed competitive scheme can place a significant downward pressure on the strike price. Also, setting a maximum bid sends a clear signal to competing developers about the extent to which cost reduction is expected from the industry at each auction round.

The government can also offer public research and development (R&D) and innovation funds and provide grants for demonstration sites with the aim to test new technologies and reduce their cost in the future. The latter, specifically, helps de-risk new technologies and improves their financeability given the conservative attitude of finance providers towards unproven technologies. Additionally, a stable policy framework along with an effective grid regulatory model that shields the project returns from any delay in grid construction through the provision of compensation can reduce the risk and improve the bankability of offshore wind projects.

Along with subsidies, the design and application of appropriate tax policies can also help lower costs. In Europe, this is mainly related to corporate tax, depreciation rules for offshore wind assets, and investment allowances. Finally, optimal site selection is another important measure through which the government can reduce the cost of offshore wind. This can be achieved by adopting a systematic approach that takes into account all information about choosing a particular site including feasibility, wind production, grid connection, seabed condition, and health and safety issues.

6.6   CONCLUSIONS

The offshore wind industry is relatively immature and this is reflected in its high cost compared with other generation technologies and its heavy reliance on government support. At the same time, the immaturity of the industry implies that there could be significant potential for cost reduction. In this chapter, we explored the issue of cost reduction and innovation in the offshore wind industry focusing on technology, supply chain management, finance, and government support policies.

There is a significant opportunity for cost reduction in the technology segment of offshore wind. This includes innovation in the design and manufacture of turbines, foundation structures, and electrical equipment as well as improvements in executing the costly activities of transport, installation, and O&M. At the same time, supply chain management through collaboration, knowledge sharing, and innovation can help to lower the cost further. Specifically, the offshore wind industry can climb up the maturity curve by identifying and implementing the best supply management practices from parallel mature industries such as offshore oil and gas.

Finance costs can also be minimised through mitigating the main risks of offshore wind projects which are construction, operation, and regulatory risks. The construction risk can be mitigated through measures such as strong contractual provision, interaction with reputable contractors, and the use of teams with expertise in the interface management. Also, the selection of contractors with knowledge of local construction along with making contingency plans for unforeseen circumstances can lower installation and logistic risks. The reliability risk of components can be reduced, to some degree, by guarantees and liquidated damage clauses. Furthermore, measures such as investing in advanced monitoring technologies and knowledge sharing along with efficiently distributing risks between parties have proven particularly useful in reducing the risks in the construction and operation phases.

The support policies of government can also lower the cost of offshore wind. These include public R&D funding and demonstration grants to develop and showcase feasibility of certain new technologies that otherwise are perceived as risky by lenders. Additionally, government policy can focus on creating a growing market for offshore wind with a clear pipeline of projects and provide incentives for cost reduction by promoting competition and setting a declining cap for the maximum bid in each auction round. Moreover, designing an efficient grid construction and connection regime that does not impact project return by construction delays and choosing the optimal site for deployment of offshore wind turbine can contribute towards achieving a cost competitive offshore wind industry.

REFERENCES

EWA. (2013). *Where's the Money Coming From? Financing Offshore Wind Farms.* European Wind Energy Association. ISBN: 978-2-930670-06-5. http://www.ewea.org/fileadmin/files/library/publications/reports/Financing_Offshore_Wind_Farms.pdf.

FTI Consulting. (2015). Innovative Financing of Offshore Wind. In *FTI Consulting, EWEA Offshore Conference, Copenhagen*, 10–12 March 2015. http://www.ewea.org/offshore2015/conference/allposters/PO219.pdf

Green, R., & Vasilakos, N. (2011). The Economics of Offshore Wind Energy. *Energy Policy*, 496–502.

IRENA. (2017). *Innovation Outlook Offshore Wind.* International Renewable Energy Agency. https://www.irena.org/DocumentDownloads/Publications/IRENA_Innovation_Outlook_Offshore_Wind_2016.pdf

Krupa, J., & Poudineh, R. (2017). *Financing Renewable Electricity in the Resource-Rich Countries of the Middle East and North Africa: A Review EL 22* (Oxford Institute for Energy Studies). https://www.oxfordenergy.org/publications/financing-renewable-electricity-middle-east-north-africa-review/

Meeus, L. (2015). Offshore Grids for Renewables: Do We Need a Particular Regulatory Framework? *Economics of Energy & Environmental Policy*, *4*(1), 85–95.

Ofgem. (2014). *Offshore Transmission OFTO Revenue Report.* London: Office of Gas and Electricity Market (Ofgem). https://www.ofgem.gov.uk/publications-and-updates/offshore-transmission-ofto-revenue-report-0.

NSCOGI. (2014). *Discussion Paper 2: Integrated Offshore Networks and the Electricity Target Model, Deliverable 3 – Final Version.* The North Sea Countries Grid Initiative. www.benelux.int/files/4514/0923/4100/Market_Arrangements_Paper_Final_Version_28_July_2014.pdf

Stentoft, J., Narasimhan, R., & Poulsen, T. (2016). Reducing Cost of Energy in the Offshore Wind Energy Industry: The Promise and Potential of Supply Chain Management. *International Journal of Energy Sector Management*, *10*(2), 151–171.

CHAPTER 7

Public Acceptance of Offshore Wind Farms

Abstract The successful deployment of offshore wind farms and other renewable technologies depends, to some degree, on their public acceptance. Past experiences have proven that public opposition can result in delays or project standstill for renewable energy projects. Therefore, it is crucial to develop a clear understanding of the social implications of offshore wind installations. This chapter reviews the main factors that give rise to opposition against offshore wind farms and discusses the ways in which public acceptance of these installations can be promoted.

Keywords Social acceptance of offshore wind • NIMBY • Public opposition • Siting of offshore wind

7.1   Introduction

Apart from technical and economic feasibility, public acceptance constitutes a critical element for the successful implementation of renewable energy technologies and projects. The term public acceptance is often defined as the positive attitude towards a particular technology or measure which can lead to supportive behaviour when needed, or alternatively will counteract opposition (POLIMP 2014). Other types of acceptance attitudes that do not lead to active support are considered as tolerance.

R. Poudineh et al., *Economics of Offshore Wind Power*,
https://doi.org/10.1007/978-3-319-66420-0_7

The social dimension of offshore wind deployment is specifically important in democratic societies. This is because the right of citizens to participate in decisions that may impact their life and well-being is ensured by law. For instance, in the European Union, public participation is guaranteed by Directive 2003/35/EC, which is based on the Aarhus Convention of 1998, to which all member states are signatories. The Directive empowers people by giving them the right to access information easily, participate in the process of decision making, and seek justice when they think their right to live in an environment adequate to their well-being is violated.

The issue of public acceptance is not a new concept. Public acceptance was traditionally a challenge in the siting of, for example, nuclear power plants, landfill, and waste disposal facilities. During the early years of renewable energy programmes in the 1980s, however, public acceptance was considered as unimportant because initial surveys showed a high level of public support, specifically for wind power technology (Wüstenhagen et al. 2007). In the subsequent years, however, it became clear that support from the general public and key stakeholders could neither be assumed nor be taken for granted.

Over time, the renewables industry has learned that not taking public opinion into account can lead to project delay or project standstill, which can ultimately threaten the viability of a project. A case in point is Atlantic Array, which was one of the largest offshore wind projects in the world and was slated to be built in Bristol Channel, United Kingdom, by RWE npower (the German utility company). The project faced fierce opposition by residents in Devon and Lundy as well as environmental and heritage groups on the grounds that it could have a lasting damage. It was cancelled in November 2013, apparently for technical and financial reasons. Similar issues with public objections over the visual impact of what was the first proposed utility-scale offshore wind farm in the United States, Cape Wind, delayed and obstructed the project to the point that it has yet to be developed, more than 15 years after its inception, and looks unlikely to come to fruition.

The social acceptance of offshore wind thus incorporates several dimensions, including both technology and the perceived impact to stakeholders in the local communities. The current economics of offshore wind is such that it relies on government subsidies to survive, thereby making the acceptance of technology a choice between short-term costs and long-term benefits. Also, like most other renewables, offshore wind turbines occupy much more space compared to conventional generation sources, and they also tend to be closer to the power demand site, thereby increasing visibility and bringing environmental impact closer to the community residence. The issues of community acceptance have long been the subject of research with a focus on the disconnect between the general acceptance of renewables at a national level and opposition to a specific project in the vicinity of the local community (Wüstenhagen et al. 2007). This difference is often explained by the argument that people tend to accept a project as long as it is not in their 'backyard'. However, this idea has been criticised for being overly simplistic and ignoring the true motives of people to oppose a specific technology.

The key issue from a policy perspective is therefore to understand the main causes of vociferous protest to offshore wind installations and identify the ways in which public opinions can be incorporated in policy decisions to resolve the conflict. This chapter deals with this issue. The next section highlights the main causes of conflict, while Sect. 7.3 then highlights approaches to addressing the social acceptance of offshore wind. Finally, Sect. 7.4 concludes.

7.2   Determinants of Public Opposition to Offshore Wind

Traditionally, the public perception of wind farms (and any other energy infrastructure) is explained through Not in My Backyard (NIMBY) (Devine-Wright 2005). The NIMBY theory assumes that everyone agrees with the usefulness of the project, but prefers to have it in someone else's 'backyard' rather than their own. This theory is also used to describe the situation in which there is general support for wind energy; however, there is a tendency to oppose a particular project in the vicinity of local residents. The empirical evidence for this theory has been inconclusive. For example, some studies have shown that people who support deployment of wind farms nationally also support it locally (Devine-Wright 2005). However, it is not clear from these studies whether the reverse is also true (i.e., whether those who oppose wind farms locally also oppose it nationally). Therefore, while NIMBY may be one of the reasons for opposition, it is often criticised for being facile, not having a robust theoretical framework, and not being able to accurately explain the root cause of opposition.

Petrova (2013) lists a number of factors that have been frequently cited in various studies as the main reasons behind public opposition to wind projects. According to Petrova, visual and seascape concerns due to the physical characteristics of turbines and aesthetic effects are among major causes of opposition. Although offshore wind farms have been given less attention compared to their onshore counterparts with respect to visual intrusiveness, in recent years much focus has been given to the issue of actual visibility from the coast. Some have even attempted to quantify the visual intrusiveness of offshore wind farms (Gee 2010). To this end, the degree of seascape change, as a result of offshore wind deployment, depends on various factors such as distance of the wind farm to the coast, topography, the number of turbines and their visible proportion, weather conditions, and navigational lighting of turbines. Of these factors, the distance between the observer and turbines has the greatest visual impact. The issue of visual impact has recently caused some countries to regulate the installation of offshore wind farms. In the United Kingdom, for example, a guideline has been published for the assessment and restriction of visual impact of offshore wind farms. In addition, the UK Energy Act of 2004 has made a provision for the decommissioning of offshore wind farms according to which developers are required to remove all turbines and associated structures at the end of their life and return the site to the condition before installation (DTI 2005).

In additional to visual impact issues, there are also socioeconomic concerns behind public opposition to wind farms. This is because the costs and benefits of a project disproportionately accrue to different stakeholders. While residents in the proximity of installations bear a high share of costs (due to impact on their life), they do not necessarily benefit more from the wind farms. The main cost concerns often pertain to the impact of the project on property values, local jobs, tourism, and other activities such as fishing, boating, or surfing. The local residents also may be sceptical about the project's significance. For instance, in Germany, on the island of Sylt, a campaign was created to oppose offshore wind farms, arguing that it severely impacts tourism by removing the very essential landscape qualities that tourists come to enjoy in the first place (Gee 2010). Various other countries that rely heavily on coastal tourism dollars may have similar objections to offshore wind farms.

The third important reason for opposition is environmental concerns of locals over the impact of the project on the flora and fauna, wildlife, marine life, and ecosystem. Offshore wind turbines are known to produce underwater noise and electromagnetic fields, which can be detrimental to marine life. Moreover, this noise is produced during the whole lifetime of the turbine from installation to decommissioning. Due to the distance from the coast, people are generally not affected by noise produced by offshore wind farms; however, marine animals can be impacted, with the extent of the impact depending on the kind of species and their ability to adjust. Overall, the effects of offshore wind turbines on animal life include bird and bat collisions; temporary or permanent habitat loss of fish and other marine animals; fragmentation of breeding, feeding, and migratory routes; disturbance of behaviour and stress; and alteration in the plant and animal community composition (OSPAR Commission 2008).

Finally, the lack of trust between local communities and project developers is a key driver of public opposition to offshore wind farms. This stems from the fact that there is the possibility of information asymmetry between locals and project developers about, for example, the real significance of the project and its long-term consequences for the life of local communities. If a developer has such private information, then it may have an incentive to withhold it unless there are regulations that mandate information sharing and transparency with the threat of penalty to be imposed if it turns out that the locals' lives are affected as a result of information withholding by a developer. This highlights the importance of transparency and information sharing, the absence of which can lead to distrust and opposition.

7.3   APPROACHES TO ADDRESS PUBLIC OPPOSITION

From a policy perspective, the important question is addressing how to avoid or reduce public opposition to offshore wind development. POLIMP (2014) identifies five categories of measures that can facilitate acceptance of renewable technologies. These are climate and technology awareness; fairness of the decision-making process; evaluation of costs, risk, and benefits of the project; local context; and trust in decision makers.

7.3.1   Climate Change and Technology Awareness

The awareness of people regarding both climate change and the technology in question can influence the acceptance of a project. Generally, the public awareness of climate change is higher in developed countries compared with developing countries. According to a survey in 2007–2008, the level of climate change awareness in developed countries including North America, Europe, and Japan is over 90%, whilst in developing countries including Africa, Middle East, and Asia, the majority have responded that they have not heard of climate change (Lee et al. 2015). Nonetheless, the results of the same survey indicate that those in developing countries that have heard of climate change perceive it as a greater risk to themselves and their families compared to their peers in developed economies.

Although climate awareness can be helpful to legitimise deployment of offshore wind turbines, it does not always guarantee an increase in public acceptance. This is primarily due to the possibility of resistance to changes in behaviour as well as climate change scepticism that lower public acceptance. A study by Spence et al. (2012) shows that the rise in climate awareness in the United Kingdom and United States has been accompanied with more scepticism and uncertainties than concerns. Even when people are both aware of climate change and concerned about its risks, they may fail to act in a sustainable manner because of, for example, the burden of behavioural change or timing mismatch between the costs and benefits of climate change mitigation measures.

In addition to climate change, there is also a need for public awareness regarding the specifications of technology including its costs, risks, benefits, and operational requirements. Focusing only on the benefits of offshore wind without mentioning the cost and possibly the risk associated with the technology will be misleading and may lead to public backlash. A balanced information supply will help the public to form a clear view about offshore wind power, and this may enhance their acceptance of the technology.

7.3.2   Fairness of the Decision-Making Process

An important determinant of the way the public evaluates a particular project is the fairness of the decision-making process. The concept of fairness can be related to the outcome (i.e., a particular project) or the process (i.e., the overall decision-making process). However, specific attention has been given to procedural fairness. This is because perception of the process as fair by the stakeholders will increase the acceptance of the outcome even if the outcome does not address all the concerns of the stakeholders (Firestone et al. 2012).

In the context of wind farms, Dütschke and Wesche (2014) present five categories of factors that cause a procedure to be considered as just. These are inclusiveness (identifying and interacting with all stakeholders in order to ensure that their concerns are taken into account); openness (sharing all relevant information with the community and stakeholders in a clear, timely, accurate, and honest way); responsiveness (listening to the concerns of the stakeholders when they are relevant to the wind farm); accountability (the continuing process of monitoring, evaluating, and sharing information about the wind farm during its life cycle); and flexibility (being prepared to amend the project design where feasible based on the request of communities). This latter can be something like re-siting of the grid connection cables.

There is also a mutual relationship between fairness and trust such that the perceived fairness is often higher when decision makers are considered as trustworthy. Likewise, when the procedure is fair and inclusive, it increases trust in decision makers.

7.3.3   Evaluation of Cost, Risk, and Benefits of Offshore Wind

The public usually do not have complete information about the cost, benefits, and risks of offshore wind and thus need to rely on the information provided by developers, government, or other interest groups. An accurate and reliable estimation of costs and benefits can build public confidence in the usefulness of the technology, at both local community and societal levels.

Another important point is cost efficiency. In order for offshore wind to win public acceptance, it needs to be able to compete with other low-carbon technologies that can provide the same service at lower costs. A high energy bill as a result of deploying an expensive technology may affect the public attitude towards that technology.

In addition, there can be an asymmetric distribution of costs and benefits because local communities often incur higher costs as a result of proximity to the deployment sites. If there are distributive issues, compensation may be required to correct for these uneven cost and benefit distributions. However, the practical application of financial compensatory schemes is not trivial because of the difficulty in estimating the exact costs and benefits of the project and the possibility that local communities perceive compensation as a bribe. This latter issue is specifically important because it can lead to distrust between communities and decision makers and act as a disincentive to accept anything that compensation is offered for. Other compensatory measures such as community benefits (e.g., transport and telecommunication infrastructure updates, tax reductions, or reduced energy bills) have proved to be more helpful in generating public acceptance of renewable projects.

7.3.4   Understanding the Local Context

There is often a divergence between public general support for renewables, which is often high in western European countries, and their opposition to a specific project in the vicinity of their property. This attitude traditionally was explained as Nimbyism, which relates such oppositions to selfishness and ignorance and thus might ignore public genuine concerns (POLIMP 2014). Nimbyism, as a way to describe public opposition, is inadequate because reaction of the public towards actual deployment of renewable facilities is often a mixture of emotions and rational concerns.

In practice, resistance out of a pure self-interested attitude and ignorance has proved to be rare. Rather, opposition is usually based on a detailed knowledge of the region, the specific context, characteristics of technology, and personal attitude. For example, a study on the public attitudes towards near-shore wind turbines in Sweden shows that opposition was motivated by not only aesthetic issues, but also by the cognitive notion of wind farms being unprofitable and inefficient (Waldo 2012). People tend to question the cost of offshore wind technology when comparing it with other alternative low-carbon sources that may provide the same service more efficiently. Specifically, people are concerned that pouring too much subsidies into the offshore wind sector may not produce value for money. These all show the complexity of people's reactions towards offshore wind farms, which can go far beyond the simple Nimbyism. Therefore, there is a need for deeper research to understand the way personal attitudes are formed and the relation between attitude and objective behaviour.

Another point is that it is a mistake to measure opposition to wind farms only in terms of active protest and assume passive locals are people who quietly support deployment of wind farms. There are often concerns and fear among the locals about the way the projects may impact their quality of life given that wind farms occupy more space and are more visible than their conventional generation counterparts. Therefore, there is always a possibility that passive locals turn into active protesters and enter into conflict with the developer and authorities. From a developer perspective, this can be impactful specifically when the project is at a stage where a significant level of investment is sunk and thus irreversible. Thus, it is crucial that emotions and fear of local communities are heard, discussed, and dealt with and their rational concerns are taken into account in the process of project design and deployment.

7.3.5   Building Trust

Trust is the key element in the siting of all types of renewable technologies especially when the aim of developers and other stakeholders is unknown to local communities (Firestone et al. 2012). In the context of offshore wind, the essence of trust is the feeling that all stakeholders will act in the best interests of local communities and public. The decision-making process is critical in this context as there is a strong interdependency between fairness of decision making, treatment of opponents, and trust. A fair process will provide each stakeholder with the same opportunity to communicate, raise concerns, and influence the decision.

Trust is also related to the confidence in the properties of the technology which is, in turn, affected by the level of awareness of the public. A familiar technology with a proven track record of performance is usually more trusted. There is also a relation between the level of public trust and the nature of organisation that develops the project. People usually tend to trust more not-for-profit organisations compared to commercially oriented companies, as the latter's promotion of technology is perceived to be out of self-interest (POLIMP 2014). Similarly, a relation exists between trust and the way a developer communicates with locals. For instance, when commercial companies focus more on environmental benefits as opposed to economic benefits in their communications with locals, they risk exposing themselves to trust issues because there is incompatibility between their message and their true business.

Furthermore, the experience of the onshore wind industry shows that building a relationship based on trust with outsiders and big energy corporations is inconvenient for local communities. On the contrary, when developers are small, locally based, community groups, they are usually more trusted and thus face less opposition compared with distant multinational corporations (Haggett 2011). This has some implications for offshore wind farm development, as they are often developed by big energy companies because of the cost and risk involved in this industry. In various surveys, locals have said that their attitude will change if the same project is developed by government rather than a private outside developer which they find difficult to trust (Haggett 2011).

In sum, trust is considered to be a very fragile feeling in the relationship between locals and the wind project developer. Contrary to the time that it takes to be built, it can break quickly.

7.4   CONCLUSIONS

The issue of public acceptance of renewable energy facilities, including offshore wind, is a major challenge in democratic societies where the right of citizens to live in an environment adequate to their health and well-being is ensured in law. Along with technical and financial parameters, public acceptance thus needs to be considered in the process of design and implementation of offshore wind projects. This chapter reviewed the main causes of public opposition to offshore wind turbines and discussed the ways in which it can be addressed.

The visual impact and seascape, which is mainly affected by the distance between observers at the shore and the turbine, is one of the main causes of public opposition. Local communities also have socioeconomic concerns due to possible impacts of the wind farm on their property values, jobs, tourism, and other activities such as fishing, boating, and surfing. Furthermore, the public is also worried about the way in which offshore wind turbines impact the flora and fauna and marine life. Moreover, the issue of trust between local communities and project developers, which results from information asymmetry and lack of transparency, severely influences people's opinions.

There are a number of measures that can improve the social acceptance of offshore wind farms. These include climate and technology awareness, decision process fairness, information provision, and building trust with public. The awareness of people with respect to climate change and the performance of the technology influence the attitude of people towards offshore wind farms. The fairness of the decision-making process boosts public acceptance even if the outcome does not address all the concerns of the stakeholder. This is because a fair process will provide all stakeholders with the same opportunity to voice their concerns and influence the decision making. Besides this, the accurate and reliable estimation of costs, benefits, and risks included in the project can help to build public confidence in the usefulness of the technology at both local community and societal levels. Furthermore, the issue of trust is a key factor. In the context of offshore wind, the essence of trust is the feeling that all stakeholders will act in the best interests of local communities and the public.

REFERENCES

Devine-Wright, P. (2005). Beyond NIMBYism: Towards an Integrated Framework for Understanding Public Perceptions of Wind Energy. *Wind Energy, 8*, 125–139.

DTI. (2005). *Guidance on the Assessment of the Impact of the Offshore Wind Farms: Seascape and Visual Impact Report.* Department of Trade and Industry (DTI). Retrieved from http://webarchive.nationalarchives.gov.uk/+/http:/www.berr.gov.uk/files/file22852.pdf

Dütschke, E., & Wesche, J. P. (2014). *Wind-Acceptance: A User Guide for Developers and Municipalities.* WISEPower deliverable 2.1. http://wisepower-project.eu/wp-content/uploads/2014_08_Deliverable_2_1_final_version.pdf

Firestone, J., Kempton, W., Lilley, M. B., & Samoteskul, K. (2012). Public Acceptance of Offshore Wind Power: Does Perceived Fairness of Process Matter? *Journal of Environmental Planning and Management, 55*(10), 1387–1402.

Gee, K. (2010). Offshore Wind Power Development as Affected by Seascape Values on the German North Sea Coast. *Land Use Policy, 27*, 185–194.

Haggett, C. (2011). Understanding Public Responses to Offshore Wind Power. *Energy Policy, 39*, 503–510.

Lee, T. M., Markowitz, E. M., Howe, P. D., Ko, C.-Y., & Leiserowitz, A. A. (2015). Predictors of Public Climate Change Awareness and Risk Perception Around the World. *Nature Climate Change, 2015*(5), 1014–1020.

OSPAR Commission. (2008). *Assessment of the Environmental Impact of Offshore Wind-Farms.* Publication Number: 385/2008, London. Retrieved from https://www.ospar.org/documents?d=7114.

Petrova, M. (2013). NIMBYism Revisited: Public Acceptance of Wind Energy in the United States. *WIREs Climate Change, 4*, 575–601. https://doi.org/10.1002/wcc.250.

POLIMP. (2014). *Acceleration of Clean Technology Deployment Within the EU: The Role of Social Acceptance.* Background Paper to the 1st Policy Brief June 2014. http://www.polimp.eu/images/1st%20Policy%20Brief/POLIMP_1st_Policy_Brief_final_background_paper_-_Public_Acceptance_-_June_2014.pdf

Spence, A., Poortinga, W., & Pidgeon, N. (2012). The Psychological Distance of Climate Change. *Risk Analysis, 32*(6), 957–972.

Waldo, Å. (2012). Offshore Wind Power in Sweden—A Qualitative Analysis of Attitudes with Particular Focus on Opponents. *Energy Policy, 41*, 692–702.

Wüstenhagen, R., Wolsink, M., & Bürer, M. J. (2007). Social Acceptance of Renewable Energy Innovation: An Introduction to the Concept. *Energy Policy, 35*, 2683–2691.

CHAPTER 8

Looking Ahead: Current Trajectory and Key Uncertainties

Abstract Europe's offshore wind industry has provided evidence of learning curve efficiencies setting in, with project economics (as measured by levelised cost of energy) drastically improving since 2015. Moving towards the end of the decade, the question becomes whether this cost curve trajectory is sustainable and if these efficiencies can be replicated in the emerging offshore wind markets of North America and Asia. This chapter summarises the current trajectory of the offshore wind industry and presents factors affecting it, including policy uncertainty in key markets and the evolution of wholesale power market prices, that could have global ramifications to the industry.

Keywords Offshore wind • Future developments • Uncertainties • Electricity prices • Brexit • Trump

8.1 INTRODUCTION

Europe's offshore wind industry has provided evidence in recent years that learning curve efficiencies are setting in across the industry. Project economics (as measured by levelised cost of energy [LCOE]) have drastically improved across each of the major European offshore wind markets since 2015 and in many cases have already surpassed 2020 LCOE targets. As

© The Author(s) 2017

R. Poudineh et al., *Economics of Offshore Wind Power*,
https://doi.org/10.1007/978-3-319-66420-0_8

detailed in previous chapters, this has been driven by a number of developments including technological innovations (particularly wind turbine generator [WTG] energy capture), supply chain efficiencies, scalability, and competitive auctions.

However, the offshore wind industry still satisfies several criteria of an immature industry, which exposes it to vulnerability. This is reflected in the limited number of players in the industry and relatively limited competition (with often less than eight major players in each industry segment), high risk, reliance on government support, and the geographical concentration of the majority of offshore wind installations in Europe. As such, it is premature to project that the offshore wind industry has achieved the foundation of sustainable cost reductions that are transferable globally. Figure 8.1 presents the current trend and key uncertainties in the offshore wind industry.

At present, the European offshore wind market remains home to nearly all of the industry's major original equipment manufacturers (OEMs), experienced project developers, contractors, and supply chain companies. Thus, developments in Europe will strongly influence how the global industry evolves in the years to come. If Europe stays on its current course, then increasing cost reductions will help to sustain investor confidence in the sector, underpinning private investment and facilitating a global expansion of the offshore wind industry and technology. If Europe falters during this early global expansion stage, however, either due to wavering policy support at home or external factors such as global commodity market trends, then the ramifications for the industry could be felt globally.

The following section will summarise the current scenario in Europe in which recent cost reductions and technological improvements have helped to attract an increasing flow of private capital and investments into the offshore wind sector and are encouraging a global proliferation of the major industry players into new markets. Section 8.3 however discusses some key uncertainties that could derail the current trajectory in Europe and elsewhere, and pose significant threats to the viability of offshore wind as a global mainstream power generation source. Section 8.4 provides concluding remarks.

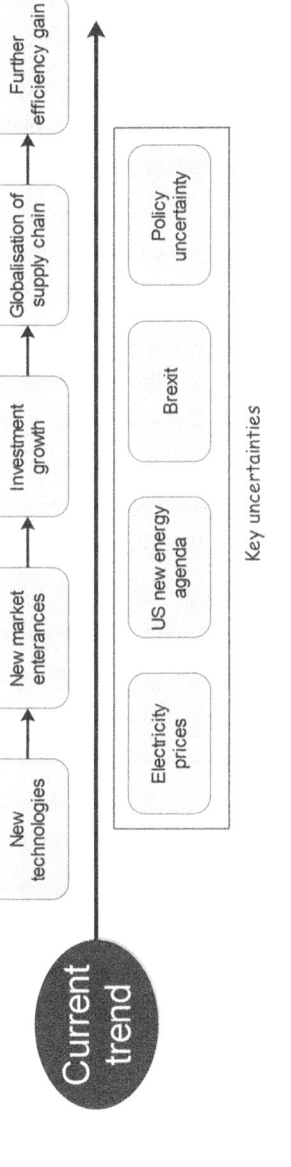

Fig. 8.1 The current trend and key uncertainties (Source: authors)

8.2   Current Trajectory of Offshore Wind Development

The cost reductions seen across the major European markets since 2015 have helped to usher in private capital investments into the sector. This has also coincided with the protracted global downturn in the offshore oil and gas projects, leaving many capital funds investors searching for new energy and infrastructure projects to add to their investment portfolios. With highly capital-intensive projects characterised by modest but stable long-term returns, offshore wind projects have thus gained significant attention as an attractive asset class amongst a broad range of money managers, institutional investors, and private equity funds in the current global context. Throughout 2016, capital investments in the sector reached nearly $30 billion globally, according to Bloomberg New Energy Finance (2017). This marked the fifth consecutive year of double-digit investment growth in offshore wind and also represented more than a 40% year-on-year increase from 2015. Notably, Europe continued to account for the lion's share of capital investment in offshore wind, representing 86% of the global investments across the sector. Figure 8.2 illustrates the growth of capital investments in new European offshore wind farms from 2010 through 2016.

The global downturn in oil and gas has also helped to court the attention of experienced offshore operators who had otherwise abstained from

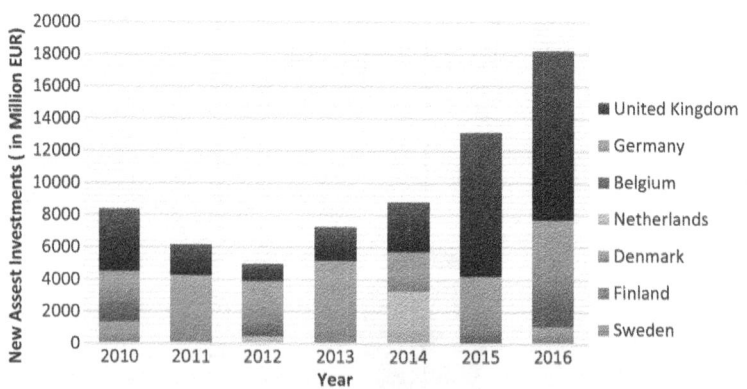

Fig. 8.2   New asset investments in offshore wind farms in Europe 2011–2016 (Source of data: Wind Europe (2017))

the industry. Until the mid-2010s, offshore wind was almost exclusively the domain of the major European utilities. Europe's major oil companies, with a few notable exceptions including Statoil and Dong Energy, largely refrained from participating in or investing in the offshore wind sector. However, a lack of viable offshore oil and gas projects in the current context, combined with increasing shareholder pressure to diversify into renewable energies, has encouraged major oil and gas companies and service providers to move into the offshore wind sector. Most notably, in November 2016, Denmark's newly privatised oil and gas conglomerate DONG energy announced plans to divest its entire oil and gas division and focus solely on renewables, anchored by its expanding portfolio of offshore wind assets. European oil and gas super-major Statoil has also taken major strides to become a leading global player in offshore wind. Royal Dutch Shell has also re-entered the offshore wind industry after a 10-year hiatus, unexpectedly winning a global competitive tender for the 700 MW Dutch Borssele 3&4 development zones in December 2016.

The entrance of major oil and gas companies into the offshore wind sector may have a number of implications for the future trajectory of the global industry. Well-capitalised and experienced oil and gas companies have the economic weight and broad market presence to apply further downward pressure on market incumbents, resulting in further cost reductions and efficiencies. Unlike smaller developers or cash-strapped utilities, major oil and gas companies have the capital and financial means to extract favourable and competitive pricing terms from major equipment providers, offshore contractors, and financiers when acting in the role of project developer. They can also aptly leverage their well-honed offshore skills and assets to compete with the largest players currently in the market, thus driving more competition into competitive bids going forward.

The presence of major oil and gas companies in the sector can also play a significant role in immature and growing offshore wind markets by encouraging industry participants to invest in the supply chain. In this way, they can help move the industry from that of what has traditionally been a fringe European industry into global energy mainstream. To date, companies such as Royal Dutch Shell and Statoil have competed in auctions for offshore wind farms both in Europe as well as in offshore development zone lease auctions on the US eastern seaboard, signalling their appetite to make offshore wind a truly global portfolio. Major European contractors are following closely behind, increasingly setting up operations in new markets such as the United States, Taiwan, and China.

Increasing private investments, more players in the market, and enhanced global competition are all positive signals for the industry. Should the current momentum and supporting conditions persist, the trajectory of the offshore wind industry over the next decade could follow along the lines below:

1. Cost of offshore wind continues to drop, driven by a combination of efficient supply chain, competitive auctions, and technological advancements.
2. New market entrants, most notably major oil and gas companies, increasingly diversify and enter the offshore wind sector, bringing new capital, infrastructure assets, and know-how, increasing competition across the industry.
3. Investments in offshore wind continue to be robust over the next several years—and as major developers seek to build out larger portfolios of projects, they open new markets up to efficient offshore wind projects.
4. Supply chain starts to globalise as project pipelines emerge outside of Europe, increasing the accessibility of cheaper components to the industry globally.
5. Developments of new technologies, such as higher-capacity WTGs, continue apace, bolstered by the entrance of new market players in the United States and Asia.
6. The need for subsidies begins to wane in some of the largest offshore wind markets.
7. Offshore wind becomes a self-sustaining industry that is economically competitive with fossil fuels and nuclear power plants.

8.3   Uncertainties to the Current Trajectory

As described in the previous section, Europe's success in improving the economics of offshore wind farms and the accompanying attention from the global financial community and major oil and gas companies have put the industry on a relatively straight trajectory towards becoming a globally viable power generation source. Despite the current inertia of the industry, however, a few external factors have the potential to derail the industry in Europe (and elsewhere), which could result in a global 'domino effect' that halts the progress of the industry in its tracks. These factors are explored in more detail below.

8.3.1   Policy Uncertainty

Although the industry has made rapid progress in weaning itself from government subsidies in Europe, project backers will still rely on some degree of non-market payments for years to come. Without these guarantees, investors will not make Final Investment Decisions on projects that are deemed too risky or uneconomical without even modest subsidy support.

Growing uncertainty about the policy appetite to support renewables across Europe beyond the year 2020, when the European Union's (EU) binding 2020 renewable energy targets end, is already a cause for concern across the renewables industry. This may be alleviated to some extent by individual National Energy and Climate Action plans due to Brussels by the end of 2017. However, in the lack of binding national renewable energy targets post 2020, governments' eagerness to allocate budgets towards supporting relatively expensive renewable energy technologies in the next decade is far from a foregone conclusion.

What hangs in the balance of this policy uncertainty is the sufficient materialisation of a project pipeline across Europe in the post-2020 era. The Financial Investment Decisions reached in 2015 and 2016 are expected to support a relatively healthy pipeline of some 5 GW of offshore wind capacity constructed by 2020 across Europe, but it is unclear how long this anticipated pipeline will continue to support investments for. According to the 2017 Cost Reduction Monitoring Framework Report, major operators are already refraining from making sufficient investments in the purpose-built assets that would be needed to construct and install the next generation of offshore wind farms in Europe (Catapult Offshore Renewable Energy 2017). More policy ambiguity would only further impair investments. The implications are not only that the current trajectory of cost reductions could halt, but the cost curve could actually reverse. For example, when the specialised installation vessels required to install large-class (heavier) WTGs and offshore wind foundation types required by project developers (such as XXL monopiles) become in short supply, it will push up equipment day rates and, in turn, drive up project construction costs. It is clear that inadequate investments today could impact costs in the future. Without long-term visibility of policy support and a viable project pipeline, industry actors will refrain from making these necessary investments to continue the industry on its current trajectory.

8.3.2   The Effect of Brexit

The 'elephant in the room' is undeniably represented by Brexit. Brexit is important for several reasons. The United Kingdom is currently the lead offshore wind market in the world and an early mover on decarbonisation efforts that has played a pivotal role in development of many of the EU energy policies. On top of this, the United Kingdom is an important player in regional approaches such as North Seas Countries Offshore Grid Initiative (NSCOGI) that may be threatened by Brexit. The UK government has also made projection about offshore wind industry to increase its local content in the supply chain and make it a centre of engineering excellence that exports overseas. However, Brexit is expected to increase the costs of international trade and this can have implications for the growth of local content of offshore supply chain.

The question of whether the United Kingdom remains in the Internal Energy Market (IEM) initiative or EU Emission Trading Scheme (EU ETS) will induce uncertainty in the future evolution of electricity prices in Europe. Market coupling arrangement is part of European legislation that requires active collaboration of interconnectors across the borders. Brexit increases uncertainty around economic viability of interconnection investment partially because of its status in the market coupling. In addition, possibility of future misalignment between the UK and the EU rules could affect the electricity market as well as the wider investment environment.

Investors are also concerned about the way Brexit affects the balance of risk and reward between the counterparties under the contract for difference (CfD). Given that the offshore wind market in the United Kingdom has historically attracted a lot of foreign investment, it is important for investors to know how joint venture will be affected by Brexit. Along the same line, European Investment Bank is crucial for investors in the UK offshore industry. However, in the future, the terms of the UK withdrawal from the EU and extent to which the investment opportunity is in line with European policies can affect access to capital from this source. Moreover, volatility in pound value and financial markets following Brexit, regulatory uncertainty and risk over restriction of access to EU Research and Development (R&D) funds, which have been instrumental to drive innovation in the offshore wind industry, muddy waters further. Finally, Brexit may affect the access of UK offshore wind industry to skilled workers.

It has been cautioned that uncertainty over policy developments at the national level are also being paired with uncertainty at the European level, which casts a shadow of doubt on every government commitment and overall confidence of foreign investors in the offshore wind sector (Navingo 2017). It is increasingly difficult to forecast what the policy environment will be in major offshore wind markets, or how 'bankable' current commitments to offshore wind in Europe will continue to be in the future political landscape.

8.3.3   The US Energy Agenda in Trump's Era

The US offshore wind industry (and the renewables sector in general) has, for years, been plagued by the stop-and-go policy environment on both the federal and state levels. Federal commitments in the form of production and investment tax credits have been erratic and are set to expire for many renewable energy technologies in the early 2020s. Recent developments have also not been encouraging. President Trump's proposed 'America First Energy Plan' promotes oil, gas, and coal but makes no mention of support for renewables. The President has also proposed budget cuts that would directly target divisions of federal agencies responsible for accelerating the development of clean energy, such as the Department of Energy's Office of Energy Efficiency and Renewable Energy division, and has vowed to dismantle the Clean Power Plan, which aims to limit carbon emissions from the electricity generation sector.

The President's planned withdrawal from the Paris Agreement furthermore undermines any confidence of strong policy support on the federal level and thus can depress investment in R&D and low-carbon technologies such as offshore wind in the United States. This decision also makes it difficult for the rest of the world to reach the goal of Paris Agreement given that the United States is a major source of finance and technology for developing countries in their effort to increase the share of renewables in their energy mix. In any case, given the prevalence of cheap natural gas, it is unlikely that the US withdrawal from the Paris Agreement will do much to save coal industry in the United States, as expected by President Trump. Nonetheless, the move can increase the uncertainty around the future of renewables.

The adverse effects of Trump's policies on offshore wind industry could be alleviated by state-level support. Since 2009, nine states in the US Northeast (Connecticut, Delaware, Maine, Maryland, Massachusetts, New

Hampshire, New York, Rhode Island, and Vermont) agreed to implement a Regional Greenhouse Gas Initiative (RGGI) programme, which seeks to reduce greenhouse gas emissions from power plants by 10% on their 2009 levels by 2018. This is to be accomplished through the implementation of a regional carbon cap and trade programme. However, a series of external events have created a persistent market overhang in carbon allowances, effectively making their pricing ineffectual. In the initial years, the combined effects of depressed economic activity, renewable portfolio standards, and low natural gas prices resulted in CO_2 emissions actually falling lower than the RGGI cap. Carbon credit prices briefly respited in 2015 as operators demanded more credits; however, these again tracked downward in 2016 as the legislation became entangled in courts (EIA 2017). There are ongoing plans to lower the CO_2 allowances in the future, which could yield successful results. However, these would likely have to be paired with technology-specific carve-out in renewable portfolio standards to effectively promote the widespread uptake of offshore wind.

There are some positive signs here. The State of Massachusetts has planned to tender some 1600 MW of offshore wind capacity that would include a power offtake agreement. The State of New York also has ambitious offshore wind capacity targets at 2400 MW, and the Long Island Power Authority has recently agreed to a 90 MW project to be located some 30 miles off the Long Island coast. Further down the eastern seaboard, the State of Maryland has awarded Offshore Renewable Energy credits (ORECs) to two offshore wind farms, which seek to be constructed by the early 2020s. President Trump's abrupt withdrawal from the Paris Agreement has furthermore motivated several individual US states to pursue policies to honour commitments to the Paris agreement via the US Climate Alliance, a bipartisan coalition formed by the states, suggesting that some state-level momentum is present that could help push US offshore wind forward.

Although there are emerging state-level commitments, these developments must also be put into recent historical context. Support on the state level for offshore wind has generally proven to be unreliable, with a history of start-and-stop policy support. What is policy today could easily reverse tomorrow. The State of New Jersey was previously one of the most outspoken advocates of offshore wind farms. However, as the political winds shifted, major offshore wind farms were left in the lurch of changing politics. While recent offshore wind leases have indeed attracted major European developers to the United States, it is far from clear that the market context will evolve to be supportive. In exchange for subsidy support on the state level, US states have imposed cumbersome local content requirements such as local jobs creations quotas onto project developers that may be impractical to achieve. It is rightfully not incumbent on the project developer to provide capital or finance to build out the local supply chain. As such, despite increasingly momentum in the US market as of late, there is still no viable supply chain, no locally based OEM manufacturing facilities, and a question mark as to how projects will get build. There are moreover layers of obstructive legislation, such as the Jones Act, that pose substantial hurdles to the development of the US market. Although the entrance of major oil and gas operators in the US offshore wind market is encouraging, they do not yet outweigh the political obstructions that must be remedied.

8.3.4   Uncertainties in the Eastern Markets

The risks which threaten the current trajectory in the offshore wind industry are not confined to Brexit and the United States. There are also uncertainties regarding the opportunities and investments in the handful of emerging offshore wind markets further to east. The People's Republic of China, South Korea, Japan, and Taiwan have all implemented direct government support mechanisms for offshore wind farms to help kick-start the industry. Each of these markets has also installed at least their first pilot offshore wind project as of 2016. China stands out as the region's growth engine, surpassing some 1.6 GW of installed capacity by 2016. Nevertheless, it is unclear whether developments in Asia will sufficiently serve as an engine for investments across the industry. With the exception of China, individual markets in Asia (Taiwan, South Korea, and Japan) are currently too small to sustain much of the foreign industry. Moreover, the growth of China's market as the regional juggernaut is entering into a state of limbo now that the government has shifted its focus from the promotion of adding renewable generation capacity to investing in power grids so that the renewables in place can deliver to their full potential (BNEF 2017). Without strong growth from the Chinese market, Asia's tepid growth elsewhere and relatively small individual markets may be insufficient to truly entice investors into the sector. Thus, it is unclear that market developments outside of Europe have the weight to support the continuing proliferation of the industry in the event that policy support dries up in Europe.

8.3.5   Direction of Power Wholesale Market Prices

The direction of wholesale markets will also keenly impact the role of off-shore wind in the future energy mix and its competitiveness with other generation sources. Should wholesale power prices remain low, it would be increasingly difficult for offshore wind to break free of subsidies and compete with low cost technologies, simply because the LCOE of off-shore wind plants would render projects uneconomical at prevailing wholesale power prices. If wholesale power markets rise, on the other hand, offshore wind may be able to more quickly break free of the need for subsidies and play an increasingly competitive role in the power generation mix.

There have been a number of global factors putting downward pressure on wholesale power markets in recent years. Generally speaking, more energy supplies and the growing use of alternative energy sources in the global energy mix have boosted supply, while at the same time increasing energy efficiencies and weak economic growth have reduced demand growth. The overhang of hydrocarbons, in many cases, has ushered in a collapse of commodity prices, making power generation costs cheaper and thus wholesale power prices lower.

In Europe, the effects of low wholesale power market prices are even further amplified by ongoing power market integration efforts. Increasing transmission grid interconnections, coupled with more efficient wholesale electricity exchange markets, are creating more liquid and efficient exchanges that are better able to match supply and demand, having a structural downward impact on the wholesale price. The incorporation of renewables into the energy mix, which are low marginal cost generation assets, also contributes to lower wholesale prices. According to a report by the European Commission (European Commission 2016), every percentage point increase in renewable share reduces the wholesale electricity price by €0.4/MWh in the EU on average, with the greatest impact (€0.6–€0.8/MWh) felt in north-western Europe. This is on top of an already decreasing price trend. Since reaching a peak of around €95/MWh in 2008, the average of wholesale power prices in Europe has since plunged nearly 70%, hovering briefly around €30/MWh in 2016. Figure 8.3 shows the recent years evolution of wholesale electricity prices in some of the major European markets.

A low wholesale power price environment may present challenges to the offshore wind industry throughout the medium and long term. Project

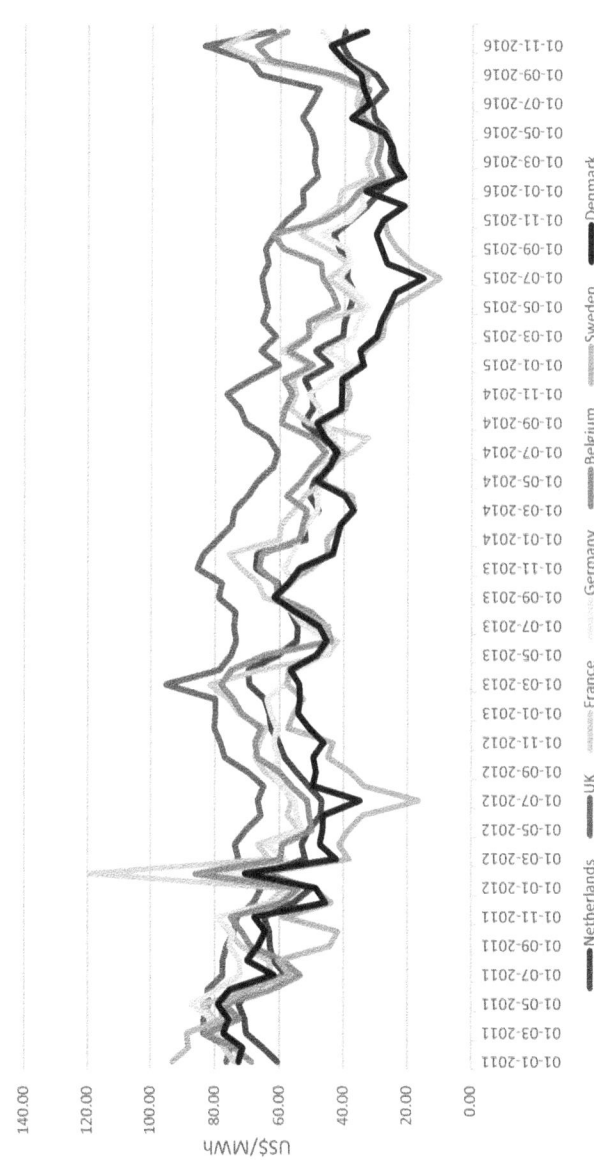

Fig. 8.3 Trend in wholesale electricity market prices in some major European countries (Source of data: compiled by authors)

economics (LCOEs) would have to improve drastically for offshore wind to be profitable at a wholesale price of, say the €30/MWh seen in 2016, without strong support from subsidies. The problem is that there is no guarantee that policy makers will be patient enough to prop up the industry until it reaches that point. Should the price of competing mainstream technologies (such as natural gas and conventional renewables) remain low, and low electricity market prices persist, then it may be increasingly difficult for offshore wind to compete over the long term and completely wean itself from subsidies.

However, there is no certainty that the current market conditions will persist. Commodities price movements such as oil, gas, and coal always play the formidable wild card. A sharp rise in gas or coal prices over the next decade (though it appears unlikely) could inflate wholesale markets. Meanwhile, a number of ageing coal plants and unpopular nuclear reactors slated to be shut down in the next decade could result in potential shortfalls in supply. This latter scenario would not only prop up wholesale prices, but further justify the offshore wind industry on the political level. Finally, not least of which, a reformed EU ETS could help to better internalise the price of electricity for polluting plants, bridging the cost gap between conventional and new technologies. Again, this would be presumably to the benefit of slightly more expensive technologies, such as offshore wind.

As such, the trajectory of the offshore wind industry will not only be influenced by policy movements and industry-specific developments (technologies, cost reductions, etc.), but also by the global energy environment that will impact commodity markets and wholesale power prices.

8.4   Conclusions

The offshore wind industry is still in its infancy. Although the industry has a growing impact within Europe's power sector, it remains relatively small on a global level, both in terms of investment level and deployed capacity.

That notwithstanding, offshore wind is currently surging and may be in the midst of something of a golden age. Recent cost reductions and technology efficiencies demonstrated in Europe have helped the industry court the attention of investors and a wider array of private companies, including global oil and gas super-majors. These developments would suggest that offshore wind, though indeed in its infancy, has the potential to become a truly global industry.

However, offshore wind projects are still reliant on out-of-market pay-ments represented by government subsidies, and this is unlikely to change anytime soon given the current direction of average wholesale power prices. This renders projects and the entire industry highly vulnerable to shifts in the political and regulatory landscape, particularly in the short and medium term. Changes that are on the horizon represented by Brexit also have the ever-present potential to disrupt the industry and further-more obstruct its growth and development globally.

The future trajectory for offshore wind depends on the interplay of a set of factors, including energy policy direction, technological progress and cost reduction, electricity price evolution, and the price of hydrocar-bon commodities, among others. If the interaction of these factors turns out to be to the advantage of the offshore wind industry, then there is a possibility that increasing competition combined with a 'wildcard' of ele-vating wholesale prices in years to come will render offshore wind into a mainstream, global industry with experienced developers transferring effi-ciencies and technologies to new markets. On the contrary, if the interac-tion of these factors happens to be to the detriment of the industry, then they can derail the industry in its current tracks and hamper the spread of experienced players and technologies to new markets.

REFERENCES

Bloomberg New Energy Finance. (2017). *Record $30bn Year for Offshore Wind but Overall Investment Down*. Bloomberg. Jan 12. Accessed online. February 10,2017.https://about.bnef.com/blg/record-30bn-year-offshore-wind-overall-investment/

Catapult Offshore Renewable Energy. (2017). *Cost Reduction Monitoring Framework 2016 (CRFM)*. Qualitative Summary Report. January 2017. http://crmfreport.com/wp-content/uploads/2017/02/CRMF-2016-Qualitative-Report-Print-Version.pdf

EIA. (2017, May 31). *Regional Greenhouse Gas Initiative Auction Prices are the Lowest Since 2014*. Energy Information Administration (EIA). https://www.eia.gov/todayinenergy/detail.php?id=31432. Accessed 28 Aug 2017.

European Commission. (2016). *Energy Prices and Costs in Europe*. Report form the Commission to the European Parliament, the Council, The European Economic and Social Committed and the Committee of the Regions. November 30. COM (2016) 769 Final. pp. 4–5.

Navingo. (2017, January 13). International Business Guide Report. *Offshore Wind*, p. 16. https://issuu.com/navingo/docs/ibg2017_totaal_spread

CHAPTER 9

The Implications for Policy

Abstract The recent success of the offshore wind industry in Europe demonstrates that policy mechanisms can be instrumental in helping the industry effectively achieve and prove cost reductions. However, the lessons learned from Europe as well as observation of the successes and difficulties in other nascent offshore wind markets around the world suggest that, when formulating an effective policy approach, policy makers must consider a myriad of factors including how to manage the integration of power to the grid, promoting system flexibility, and power market design. Accordingly, policy approaches may vary market by market and depend, to a great extent, on the current stage of the market's development.

Keywords Offshore wind policy • Support scheme design • Market creation • Operational challenges

9.1 Introduction

Government policy interacts with the offshore wind sector at two levels. First, at the higher level, energy and economic policies that specify the road map for decarbonisation, security of energy supply, economic growth, and industrial strategy can be designed with a strategic view to the offshore wind sector. The lower-level policies, on the other hand, include designing efficient support schemes, creating a market for offshore wind,

© The Author(s) 2017
R. Poudineh et al., *Economics of Offshore Wind Power*,
https://doi.org/10.1007/978-3-319-66420-0_9

incentivising cost reduction, and dealing with operational challenges. The role of lower-level policies is to establish the offshore wind sector and make sure that it becomes a sustainable industry in the future. The importance of higher level policies, however, is that they provide legitimacy to the support of offshore wind and position the sector as an integral part of a country's economic and energy policy objectives.

The legitimacy of supporting the offshore wind industry varies based on the type of economy and the government's objectives. In Europe, for example, where decarbonisation has been pursued as a policy priority, offshore wind has been considered a key player. Alternatively, in countries where there is lack of sufficient indigenous fuel source for power generation, offshore wind is a key part of policies to ensure security of supply. Offshore wind is also considered instrumental in realising broader policies of economic growth, job creation, balance of trade, and industrial strategy. This latter is not only important in developing countries, but also in developed economies which are now facing fierce competitive challenges, as two-thirds of the world industrialises rapidly. The offshore wind sector therefore provides an opportunity for governments to build manufacturing and service industries around this generation technology, create jobs, improve the industrial capability of the country, and pave the way for gaining a share of the global market.

As supporting offshore wind is costly to rate payers and tax payers, the issue of cost is at the heart of offshore wind policies. This means governments are keen on implementing policies that expedite the process of achieving grid parity. Specifically, the role of market mechanism and competition as a tool to achieve cost competitiveness is of interest to policy makers. The recent success of the offshore wind industry in Europe, where competitive auctions for subsidies have converged on parity with wholesale market prices, demonstrates that competitive policy mechanisms can be instrumental in helping the industry effectively achieve cost reductions. However, markets outside of Europe have, to date, struggled to effectively set a firm foundation for the industry. This is demonstrated not only by the lacklustre progress of the industry to meet government deployment targets, but also a lack of private sector investment in the specialised supply chain and Original Equipment Manufacturing facilities in markets outside of Europe. The main question thus, from a policy perspective, is how to create a vibrant and sustainable offshore wind market in different contexts. This chapter discusses several of the implications for the policy design focusing on support scheme design, market creation, cost reduction, and managing operational challenges.

The outline of this chapter is as follows. The following section addresses the approaches to creating a vibrant offshore wind market. It argues that markets in different stages of development will require different policy approaches to facilitate a viable offshore wind industry. Section 9.3 then discusses policy design to promote industry-wide cost reductions with the objective of alleviating the need for government subsidies for the industry. Section 9.4 analyses the operational and technical challenges of integrating offshore wind power that should be considered and rectified by policy design. Section 9.5 provides concluding remarks.

9.2   Creating a Vibrant Offshore Wind Market

To lay the foundation for a vibrant offshore wind industry, the policy maker must assess the compatibility of the policy scheme with the stage of maturity of the technology and industry. Policies to promote size and scalability of the industry are generally more effective in the nascent stage of a market's development because they help to usher in a wider array of actors and investments across the industry that can create the foundation for cost reduction via competition and economies of scale. However, in a maturing market that already boasts sufficient competition, scale, and private investments in technological development, competitive policies are likely to be more suitable to help the industry alleviate the need for subsidies (Brown et al. 2015).

In most regions of the world outside of Europe, offshore wind remains an immature technology and the market for offshore wind is only in an early development phase with few commercial actors and limited supply chains. The priority of initial policy design must therefore be to cultivate and foster certainty in the size (volume potential) of the market in order to court the attention of private investors. As early as 2008, the National Renewable Energy Action plans in Europe created clear visibility in market size for the industry. This signalled to private sector actors that there would be long-term government support behind the industry. This is also important from an efficiency point of view, as there is a relationship between the cost and the scale of the industry.

The repercussions of not implementing clear volume targets, by contrast, have been felt for over a decade in the United States. Without clear visibility of the potential US market size, private actors have refrained from investing in the original equipment technology or the purpose-built facilities such as large steel fabrication yards or specialised installation vessels necessary to support development of the market. A lack of government targets can be easily perceived as a lack of government support behind the technology and market development (this has just subsequently been addressed by the state-level targets in Massachusetts and New York, which hope to catalyse investments).

Even with clear volume targets, however, rational investors will not invest capital until there is a sufficient pipeline of projects in place. A sufficient pipeline of projects can only be realised when a policy support mechanism is implemented that is deemed 'bankable' by the private investors, or in other words, the policy mechanism is reliable and allows for an adequate return on investment. This is particularly important when the technology in question is at an early stage of development. According to Ederer (2016), the foundation for growth of the UK offshore wind market was a reliable policy that offered enough incentive in the form of profit and certainty so that investing in immature technology become attractive to investors. In many of the European offshore wind markets, 15- to 20-year fixed feed in tariffs were utilised in the nascent stages of the offshore wind market's development. One of the key advantages of this mechanism, as detailed in Chap. 5, is that the revenue stream stability and certainty not only helps more projects become economically viable (and bankable to investors), but also minimises investment risks and helps reduce financing costs.

While attracting and maintaining private investments is achieved largely through visibility of the market and implementing bankable support mechanisms, policy design must also address social opposition to projects. The ramifications of not addressing opposition should not be underestimated. As previously mentioned in Chap. 7, Cape Wind, the first proposed utility-scale offshore wind farm in the United States, has languished for over a decade as the developers battled formidable social opposition from residents of Cape Cod, from where the turbines would be visible, and various other local stakeholders. Apart from delaying the Cape Wind project (and perhaps burying it), the social opposition also planted seeds of doubt in the viability of the entire US offshore wind market. Combined with the lack of any government targets or stable subsidy support scheme to make investments bankable, private investment in the US offshore wind sector was widely deemed as too risky.

Thus, creating clear volume targets, introducing sufficient and reliable support schemes, and addressing public opposition are some of the main factors that can help set the foundation for a vibrant offshore wind market. This is particularly true when both the technology and the market are in an early stage of development. However, as these initial policy mechanisms nurture the industry from an emerging stage of development into a growth phase, the policy maker will eventually confront a dilemma as the industry starts to grow. On the one hand, government subsidies must offer sufficient incentive and guarantees in terms of remuneration so that they effectively attract and sustain sufficient private investment across the industry. On the other hand, however, the policy must be designed in such a way that it keeps margins reasonably low for the industry, because over-generously subsidies would discourage efficiencies in the marketplace and increase the cost of support policies (Ederer 2016).

As such, the policy approach must consider, and should be shaped by, the stage of technology's development and of the market itself. In early stages of development, policy makers must focus on building markets for offshore wind and encouraging investments in the technology. This is achieved primarily by setting clear volume targets and implementing bankable policy mechanisms that attract private investment into the sector, while also effectively addressing social opposition. As described in the next section, however, as the market matures, the focus of the policy design must shift to reducing costs of the technology, integrating the offshore wind power into the grid, and effectively weaning the industry from policy support.

9.3   THE EXIT STRATEGY: OFFSHORE WIND WITHOUT SUBSIDIES

In order to survive in the long run and become a mainstream renewable industry, the offshore wind industry must ultimately wean itself from government subsidies. For this to happen, two factors play important roles: reduction in the cost of offshore wind and rise in the wholesale electricity prices.

As detailed in Chaps. 3 and 6, the cost of offshore wind energy is comprised of a combination of technology (including turbine, foundation, electrical equipment, installations, and O&M), supply chain, and financing costs. Well-designed policy can and should play a critical role in lowering cost across each of these areas. Provision of innovation incentives and supply chain integration can lower the cost of technology. At national and sub-national levels, the responsibility to engage in technological innovation is distributed across many stakeholders with diverse interests and objectives. An effective innovation policy for offshore wind requires aligning the companies' strategies with broader government objectives in the energy sector and expanding the network of stakeholders beyond the renewable industry.

Competitive procurement policies which allow only the most efficient and innovative developers to proceed to market have proven particularly effective in incentivising cost-reducing innovation. The importance of introducing more market mechanisms in the renewable industry is now widely recognised. In Europe, for example, the European Commission State Aid Guidelines for Environmental Protection and Energy 2014–2020 states that, with few exceptions, all member states that seek state aid clearance need to introduce competitive processes for allocation of renewable subsidies (European Commission 2014). However, the design features of the applied auction are very important. In the context of offshore wind power procurement, specifications such as auction format, frequency, lead time, pricing rules, penalty for non-delivery, pre-qualification criteria, and project size requirements are determinants. In order to signal the expectation about future cost reduction in the industry, the government can set a target for the price of offshore wind along with a regressive cap for the maximum bid in the auctions. In a nutshell, the government needs to ensure that the auction design and procurement process are adequate to meet its objectives.

In addition, policies that aim to enhance collaborations across supply chain are crucial. The collaborative innovation efforts save on transaction costs, enhance gains from economies of scale and scope, create synergy, provide access to complementary resources, and decrease Research and Development (R&D) costs of the offshore wind sector. These are specifically important in the context of offshore wind which is an immature industry and has significant potential for cost reduction. Building an efficient and effective supply chain will help to lower the risk and the cost of industry and improve the investment climate.

Along with the aforementioned factors, the grid connection costs also matter a great deal. These costs can be transferred to project developers to be factored into their bid or to the transmission network operator to be recovered through tariffs. An advantage of the generator pay model is that the project developer will have a strong incentive to implement the most efficient grid connection model. The disadvantage may be that it raises the costs of the project and make it uncompetitive. In places that have had the lowest offshore wind price (such as Denmark and The Netherlands), the grid connection costs are often assumed by transmission network. Alternatively, the UK offshore transmission model tries to benefit from the best specification of both models. It grants licence to a third party, through a competitive tender, to operate the new offshore transmission asset. This implies that generators should, theoretically, be partnered with the most efficient and competitive bidders in the market. In nascent and non-liberalised electricity markets, perhaps, the transmission system operator (TSO) model is the easiest method because of its simplicity and existence of relevant institutions. In mature electricity markets, however, more complex approaches such as that of the UK offshore transmission model can be adopted.

At any stage of the technology or market's development, the government policy needs to focus on de-risking the offshore wind sector. As the industry moves towards more sophisticated project financing, the presence of a sustainable stream of revenue along with an appropriate grid connection model is indispensable. Furthermore, the government needs to ensure that the process of permitting, seabed leasing, and grid connection goes smoothly and without delay as uncertainties in these will affect the cost of capital. This amounts to streamlining the process to pave a clear path for project developers. In addition to this, given that, in many liberalised electricity markets, subsidies are linked to the wholesale electricity market, regulation needs to ensure that generators do not face any volume risk and the amount of price risk is contained within a narrow range. Such policies and regulatory measures reduce the systemic as well as project-specific risks and can lower the margin charged by debt providers. In the emergent offshore wind markets, where this industry is perceived as risky by lenders, the government can perform auxiliary measures to improve the financeability of projects. These, for example, could include establishing Green Bank and revolving credit facilities and offering project finance loan guarantees, credit guarantees, or other instruments to improve creditworthiness of project developers.

Along with cost reductions, an increase in the price of electricity is another factor that can boost the economics of offshore wind and eventually reduce or obviate the need for subsidy. There are three policy tools that play key roles in the price of electricity: carbon pricing, redesign of electricity market, and direct regulation to restrict unabated coal and/or nuclear power plants. For instance, if emission trading schemes do not create a sufficiently strong price signal, the government can set a carbon price floor. The UK carbon price floor has effectively resulted in a higher price for electricity in that country compared with its European neighbours.

Likewise, redesign of the electricity market is important given that the penetration of zero marginal cost renewables has depressed prices in the European energy-only electricity markets. Some proposed measures to modify energy-only markets include ensuring the cost of scarcity and flexibility is genuinely reflected in the electricity price. In the United Kingdom, for example, recent changes aimed at dealing with these issues include setting scarcity price based on the bid from the marginal plant and removing the price cap and dual price mechanism from the balancing market.

Finally, any measure to retire nuclear power plants and unabated coal can create a supply gap with possible implications for electricity prices. Since the Fukushima nuclear disaster in 2011, some governments, like Germany, have shut down parts of their nuclear fleet and have planned for their eventual phase-out, while many others have banned construction of new nuclear plans. Also, in 2017 the EU region energy utilities (except Poland and Greece) announced that they will not build new coal power plants after 2020. The supply gap, as a result of these measures, can raise the electricity prices and provide an opportunity for renewables, including offshore wind, to increase their share in the generation mix.

The main concern for policy makers, however, is if and when offshore wind can become a subsidy-free sector. Recent developments, specifically, the offshore wind auctions in Germany and the UK have raised the hope that the industry can achieve cost competitiveness and do away with subsidies even sooner than anticipated. The German projects in particular, if realised, will be the first offshore wind farms in the world that are developed entirely based on electricity market prices and without any subsidy.

The German auction results raised questions about the main drivers of the low bids and whether or not this is what policy makers should expect from offshore wind auctions in the future. According to NERA Economic Consulting (2017), a number of factors have contributed to achieving such a success in the German offshore wind auction. First, these projects are not only in sites that benefit from strong wind resources, but are also close to other wind farms owned by developer companies. Such proximity provides significant scale effect through shared grid connection and operation and maintenance costs. Second, the long lead time of projects has enabled developer companies to factor in expectations about future cost reduction due to technological advancement. Third, life extensions of up to 30 years for offshore wind farms by German authorities allow the costs to be spread over a longer duration. Fourth, developers expect an increase in the electricity prices during the operation life of these projects given that nuclear plants and many of the polluting conventional generators are going to be phased out by 2022. Fifth, the presence of fierce competition in the auction put a downward pressure in the price. This competition was intensified by the fact that unsuccessful bidders had only one more chance (in 2018) to secure a contract before their projects become ineligible for grid connection. Sixth, grid connection costs in the German model are assumed by the TSO, and this lowers the cost to project developers.

Apart from the aforementioned factors, there is also another hypothesis which argues that the main reason behind the very low bid in the German offshore market is the auction design (NERA Economic Consulting 2017). In the German auction model for offshore wind, bidders are required to post a bid bond of around €100/kW. At the same time, there is a long lead time (multi-year) and penalty for not delivering the project is as low as 30% of the bid bond. This design creates a real option value for the successful bidder in the sense that, by bidding for zero subsidy, the project developer buys an option which can become profitable if the expected market and technological development are realised. The long lead time increases the gain from waiting, whereas the large scale of the project increases the gain from electricity price increase and technological advancement. In other words, the developer can make a low bet on his ability to deliver in the future.

The result of the 2017 UK offshore wind contract for difference auctions was also surprising. Within the period between first- and second-round CfD auctions (2014–17), the price of offshore wind more than halved. Several reasons can be mentioned to explain these low prices. First, the construction techniques and turbine technology are improving and thus today's offshore wind projects are less risky compared to earlier projects. This means the hurdle rates of returns for offshore wind have fallen in the past few years from two-digit numbers down to one-digit numbers. Second, there is currently a secondary market for offshore wind assets, which means developers have a clear exit route to monetise their investments that are backed by CfD contracts.

However, similar to the case of Germany there is a loose delivery requirement for CfD contracts in the UK. This means developers can withdraw with only a limited penalty if they are unable to obtain the necessary financing. Therefore, developers can see these auctions as "options" which could turn into a profitable opportunity if direction of market and technology develop in their favour.

Although the German and British offshore wind auctions may not be replicated elsewhere in the short run, they show that, indeed, an offshore wind industry without subsidy is possible if both the market and technology develop in the desired direction. The implication for policy makers is that they should not abandon supporting the offshore wind industry, but instead focus on incentivising cost reduction and correcting electricity price signals. Once the industry becomes competitive and subsidies are removed, the projects can be structured around electricity market prices. At that point, the offshore wind industry can learn from the experience of the oil and gas sectors where, traditionally, projects have been developed around commodity market risks.

9.4   Addressing Operational and Technical Challenges

Electricity supply and demand has to remain in continuous balance. The necessity of this balance stems from the technical requirement for grid frequency which must not deviate beyond its statutory limit. The key operational feature of offshore wind power is that it is intermittent. The challenges that intermittent wind generation impose on the power system are related to voltage fluctuations, reactive power, frequency control, harmonic disturbances, and inertia, among others. In addition to being technical hurdles, dealing with these constraints is often costly as the system operator needs to make additional arrangements. Therefore, if the penetration of intermittent resources including offshore wind is to increase in the generation mix, regulators need to ensure that the power system has a sufficient level of flexibility.

The flexibility of the power system depends on several factors such as the sources of flexibility, market design, and the condition of the electricity network. The main sources of flexibility are conventional dispatchable generations (e.g., gas power plants), storage, interconnection, and demand response for which utilisation is gaining momentum in recent years. The role of market design is not only to ensure that the existing flexible resources have incentive to offer their services in the market, but also to incentivise new investment in flexibility services. The condition of the grid is also important for dealing with intermittency as lack of an adequate transmission and distribution grid severely constrain the amount of available flexibility in the system.

Although wind power has traditionally been considered an inflexible generation source, in recent years, the technological advancement in power electronics and turbine design have enabled this generation technology to be more controllable. As a result, the operational requirements of offshore wind are gradually changing. Unlike the old industry codes, the new grid codes specify the steady and dynamic requirements that wind turbines must meet in order to become connected to the grid. These requirements, for example, include capabilities of wind farms to contribute to frequency and voltage control by continuous modulation of active and reactive power supplied to the grid. Another example of requirement change can be found in Europe, where power system operators are gradually moving towards transferring balancing obligations to intermittent renewables (including offshore wind) and treating them as conventional generation sources.

In the absence of a flexible power system, offshore wind farms may face frequent curtailment with consequences for the revenue of these generators. The key point, from a policy perspective, is that, along with penetration of offshore wind in the generation mix, the power systems need to become more flexible. This can be done, for example, by the introduction of specific incentives for flexibility in the electricity market. Moreover, policy makers need to ensure that offshore grid operators have sufficient incentive to maintain the reliability of subsea cables. This is because subsea cable failures cause more financial losses to the offshore wind generators than any other element. One way to do this, for instance, is for regulators to make the revenue of grid operators contingent upon achieving a certain level of reliability.

9.5   CONCLUSIONS

Building a vibrant offshore wind sector requires implementation of appropriate policies. By setting clear targets and positioning offshore wind as part of the country's integral energy and economic policies, the government will send a signal to the industry that it has real commitment to this sector. The incentive for investment must be based on the stage of technology maturity, compatibility with the market structure and institutions, as well as the degree of risk coverage. In nascent markets, implementation of support policies that cover all economic risks and promote scalability is most likely to be effective. As the industry grows, the government can then cap the volume or budget for subsidies and introduce market mechanism through auctions in order to incentivise cost reduction and control the cost to consumers. In the non-liberalised electricity markets, a power purchase agreement (PPA) is crucial to provide a guaranteed offtake. It is also important to recognise that the support scheme needs to allow for the full integration of offshore wind within the liberalised markets.

The key objective of policy is to achieve a self-sustainable offshore wind industry which can compete with other sources of generations without relying on government support. The recent trend in the sector shows that this is indeed possible. The role of government is to employ cost-reducing incentive mechanisms, encourage supply chain integration, and promote R&D. The introduction of a carbon price mechanism and phasing out nuclear and unabated coal can also speed up the process of removing subsidies for offshore wind power in that it influences the prevailing wholesale price of power.

Finally, it is important that penetration of offshore wind is not hindered by a lack of flexibility in the power system. The government must ensure that the design of the power market provides sufficient incentive for investments and operation of flexibility services. In addition, transmission networks should be able to easily accommodate the power from wind generators and the offshore grid operator needs to be incentivised to provide an acceptable level of reliability for operation of subsea cables. In doing so, the policy maker can effectively lay the foundation for a self-sufficient offshore wind sector that fulfils a series of broader policy objectives.

REFERENCES

Brown, C., Poudineh, R., & Foley, B. (2015). *Achieving a Cost-Competitive Offshore Wind Industry: What Is the Most Effective Policy Framework?* Oxford Institute for Energy Studies. https://www.oxfordenergy.org/wpcms/wp-content/uploads/2015/09/EL-15.pdf

Ederer, N. (2016). The Price of Rapid Offshore Wind Expansion in the UK: Implications of a Profitability Assessment. *Renewable Energy, 92*(2016), 357–365.

European Commission. (2014). *Guidelines on State Aid for Environmental Protection and Energy 2014–2020.* Available from: http://eur-lex.europa.eu/legal-content/EN/TXT/PDF/?uri=CELEX:52014XC0628(01)&from=EN

NERA Economic Consulting. (2017). *Method or Madness: Insights from Germany's Record-Breaking Offshore Wind Auction and Its Implications for Future Auctions.* http://www.nera.com/content/dam/nera/publications/2017/PUB_Offshore_EMI_A4_0417.pdf

CHAPTER 10

Conclusions

Abstract The offshore wind industry is approaching three decades of existence. Although the industry has shown significant achievements in cost reductions in recent years, it is poised to be a policy-driven industry for the foreseeable future, particularly in nascent markets outside of Europe. The key factors to the growth and maturity of the offshore wind industry are reaching grid parity and phasing out subsidies, dealing with issues of wind power intermittency through power system flexibility and improving public acceptance. Therefore, it is important that policies to promote offshore wind encompass a wide range of considerations.

Keywords Offshore wind power • Policy design • Competition-based support policies

10.1   AN INDUSTRY WITH THREE DECADES OF EXPERIENCE

The offshore wind industry, as a continuation of land-based wind power generation, started around 30 years ago with the deployment of the first commercial offshore wind farm in the Danish waters. Despite the dominant view throughout the 1990s that offshore wind energy was prohibitively expensive and the technology would not become economical until

© The Author(s) 2017   151
R. Poudineh et al., *Economics of Offshore Wind Power*,
https://doi.org/10.1007/978-3-319-66420-0_10

after 2020, the industry, in Europe, has shown significant progress in realising cost reductions, specifically, in recent years.

The strong development of Europe's offshore wind markets has rendered Europe the de facto leader in this industry in terms of the share of installed capacity. It is also the global hub for turbine technology and manufacturing and houses the leading purpose-built installation vessels, technology providers, fabrication and installation experience, and know-how. Outside of Europe, the industry is still nascent but has gained significant momentum in recent years. Notwithstanding these developments, the major cost drivers associated with the offshore construction and operation of offshore wind farms remain formidable. Thus, given the high cost of the power generated from the wind farms relative to competing technologies and the prevailing wholesale power prices, offshore wind power is poised to be a policy-driven industry for the foreseeable future, particularly in markets outside of Europe where the supply chain and technology are in an earlier stage of development.

The justification for policy support for offshore wind generally revolves around decarbonisation of the economy, meeting renewable energy and carbon abatement targets, bolstering national energy security, and supporting industrialisation and jobs growth. Different markets may place a different emphasis on each of these areas. Policy makers in developed economies, for example, are more likely to emphasise carbon abatement, whereas policy makers in emerging and developing economies may prioritise energy security or improving urban air quality. In Europe, which boasts nearly 90% of installed offshore wind capacity, offshore wind energy has been supported primarily for its role in carbon abatement and its contribution to helping countries achieve their binding 2020 renewable energy targets under the EU regulation.

In any case, offshore wind is not the only pathway to fulfilling these policy objectives. However, offshore wind power can provide unique advantages to the policy makers, namely in that the technology makes use of a sea of otherwise untapped energy resources adjacent to major coastal demand centres, averts land-use conflicts, and is scalable to utility-scale power generation. It can also form the basis for jobs creation and supporting local heavy industries. Nevertheless, there are also inherent and unique cost drivers of installing and operating wind farms in the harsh marine environment that must be considered.

10.2   An Effective Policy Framework

As such, an effective policy design for the support of offshore wind industry must encompass a wide range of considerations. First and foremost, it is incumbent upon the policy maker to implement an efficient policy that achieves its objectives in the most cost-effective way. The policy design must also be effective, in that it is reliable and provides sufficient incentives in terms of return on investment to encourage private investment in the sector and technology. Moreover, it must be equitable in its distribution of the costs of the policy.

To date, policy makers have attempted to utilise both indirect and direct policies to incentivise the deployment of offshore wind power, with varying results. Indirect policies that effectively internalise the social costs of carbon into wholesale power markets via pricing carbon can, theoretically, support the commercial viability of more expensive technologies, such as offshore wind, by escalating the prevailing wholesale price of electricity. In practice, however, if the cost of carbon is too low, then it will fail to either incentivise a switch in generation technologies or support more expensive technologies. This has largely been the experience with the current European Emission Trading Scheme (ETS).

Direct policy support mechanisms have proven more effective in developing an offshore wind industry and, more recently, encouraging price reductions in developed offshore wind markets. Direct policy mechanisms typically consist of a combination of production-based schemes such as feed in tariffs and feed in premiums, quantity-based schemes such as renewable obligations, and investment-based schemes, such as an investment tax credit.

The decision over the most effective policy framework to employ will vary market to market and depends, to a great extent, on the current stage of the market's development and several aspects of the electricity market structure. Feed in tariffs and feed in premiums, for example, must be designed to remunerate within a tight bandwidth of the prevailing wholesale market price in liberalised markets, lest it risk becoming an inefficient or ineffective policy design by either overcompensating or undercompensating investors. Grid connection and power integration must also be considered in order to prevent debilitating bottlenecks or uneconomical power curtailments that dissuade investments.

A survey of developments across the global offshore wind markets over the past two decades suggests that direct policies implemented to promote scalability will be the most effective in nascent offshore wind markets. This is because both the technology and the supply chain are in an early growth stage and require vast amounts of private investment. This means that an effective policy support scheme must be both reliable and bankable so that it sufficiently covers investors' risks and provides adequate returns to attract private investments in the industries and technologies. This will facilitate a pipeline of projects that encourages further growth and investments across the industry. Establishing national targets for offshore wind deployment capacity and establishing market visibility are indispensable in the policy maker's toolkit. These all serve to install confidence and attract initial investments in the sector.

Nevertheless, the upfront cost of these policies can be high, and, moreover, do not necessarily incentivise or reward the most efficient projects and technologies in the market. Therefore, as the market grows, the policy scheme must shift its focus to mechanisms that will effectively promote competition in the market place. This has recently been the case in Europe, where 'reverse auctions' and contract for difference (CfD) policy frameworks implemented since 2015 have resulted in steep reductions in the cost of subsidies to offshore wind farms by awarding only the most competitive projects with subsides.

As a direct result of competitive subsidy auctions, recent bid prices by project developers in Germany and Denmark have indicated that the development of offshore wind farms in select European markets may be done so subsidy free by the mid-2020s. This would indeed be a tremendous accomplishment for the offshore wind industry and fulfil the policy makers' end-goal. However, it is still unclear whether this will be achieved by the mid-2020s. It has been suggested that a consequence of the current auction format in Germany, whereby bidders of offshore wind tenders stake a relatively low bond price for a project with a long development lead time, may essentially just be buying an option to develop the wind farm in the future. Whether the wind farm will ever be developed without the support of subsidies will greatly depend on the evolution of wholesale power markets in Europe and reduction in the cost of technology over the next few years. If the wholesale power prices are too low then the wind farm will likely not be developed, as it would be done so at a loss.

That notwithstanding, the implementation of competitive auctions for offshore wind farms subsidies in Europe, following on the heels of promoting size and scalability of the industry, has resulted in significant improvements to the levelised cost of energy (LCOE) of offshore wind farms. Competitive policies incentivise cost reduction through innovation in the design and manufacturing of turbines, foundation structures, and electrical equipment as well as improvements in executing the costly activities of transport, installation, and operation and maintenance. These innovations and improvements greatly reduce the costs of subsidising technology, while also helping to decarbonise the economy, create jobs, and support industry.

Outside of Europe, however, many markets have struggled with implementing effective strategies conducive to ushering in investments in offshore wind. In Asia, the People's Republic of China thrived in installing offshore wind capacity from 2010 to 2015. This success was based largely on ambitious targets for offshore wind capacity (market visibility), bankable feed in tariffs for offshore wind, and the involvement of state-owned industries in the supply chain. The market has faltered as of late, however, due to obstructions in market visibility and uncertain policy support going forward, as policy makers necessarily must now divert their attention to curtailment issues of the power.

Further east, Taiwan, Japan, and South Korea have been slow to launch their industries and are lagging government-established targets for offshore wind deployment. Despite the implementation of direct policy support mechanisms such as feed in tariffs in these countries, the relatively small size of the individual markets has discouraged both foreign and domestic investors from investing in a localised supply chain. This will likely remain the case until there is a reliable regional project pipeline that would allow operators to serve the entire East Asian offshore wind market.

In the United States, meanwhile, an absence of indirect policy support and a history of unreliable direct policy support mechanisms in the form of on-again, off-again tax incentives have been unconducive to the development of offshore wind projects. This is particularly true given the long lead times for projects that must rely on long-term certainty of policy support mechanisms. Furthermore, President Trump's withdrawal from Paris Agreement along with his promise to dismantle Clean Power Plan induces further uncertainty in the US market.

Nevertheless, there are encouraging policy developments on the state level that include quotas for offshore wind power. However, until there are reliable, bankable support mechanisms in the United States, it is unlikely that a sufficient pipeline of projects will develop that can sufficiently attract investments from private actors in the US supply chain. Until this happens, the US market is likely to be slow to develop, despite some recent progress on the individual state level.

The current challenges in the East Asian and US markets moreover highlight the necessity of implementing policy mechanisms that spur the growth of a bespoke supply chain. Given the high costs of construction and installation of offshore wind farms, a well-developed supply chain is crucial to supporting the development of an offshore wind industry and facilitating eventual cost reductions through scalability and competition. The lessons from successful policies in Europe suggest that the most effective approach to nurturing the offshore wind industry from a nascent industry to one that eventually meets its intended policy objectives without the need for subsidies is based largely on policies that initially promote a robust project pipeline. This pipeline will encourage investments across the sector, from the supply chain to technological development. Only once this happens can the policy maker divert attention to competition-based policy mechanisms that can effectively wean the offshore wind industry from subsidies.

The cost of offshore wind can also be pushed down further by supply chain management through collaboration, knowledge sharing, and innovation. Specially, there might be opportunities to identify and implement the best supply management practices from parallel mature industries such as offshore oil and gas. Furthermore, the capital costs can be lowered by providing access to risk mitigation instruments for all types of risk to which offshore wind is exposed to including construction, operation, and regulatory risks. The regulatory model of offshore grid connection is also important for the cost of project. If grid interconnection costs are to be paid by generators, they need to be able to factor them in their bid. In nascent and non-liberalised electricity markets, perhaps, the transmission system operator (TSO) model is the most straightforward method because of its simplicity and existence of relevant institutions. However, it can be inefficient given the absence of competition in this model. In mature electricity markets, however, more complex approaches such as that of the UK offshore transmission model can be adopted to utilise the power of competition and third-party knowledge and capital.

Another important policy question is that of how should the cost of support policies be recovered until renewables achieve full grid parity. In many European countries, the support for renewable power has been paid for mainly through levies on electricity tariffs. In the United States and Canada, however, governments have paid these costs through the general government budget. The method of recovering costs of decarbonisation policies has important implications not only for electricity prices, but also for competition among different energy vectors. The current European approach is distorting competition in final energy markets and raising the cost of decarbonisation. Decarbonisation of the power sector, and the economy as a whole, is a public good. Optimal taxation theory justifies the financing of public goods through general taxation, rather than by raising the price of a specific commodity like electricity. Financing public goods through general taxation would improve efficiency, lower the price of electricity, and align energy sector fiscal policy with decarbonisation strategy based on electrification.

The social dimension of offshore wind deployment is also important as public opposition can result in project delays or standstill with consequences for the cost of a project. By improving climate and technology awareness, along with the implementation of regulations that emphasise decision process fairness, information provision, and trust building with the public, the policy can improve public acceptance of offshore wind installations. Furthermore, it is important that penetration of offshore wind is not hindered by a lack of flexibility in the power system. The government must ensure that the electricity market design is adequate, the transmission network is able to easily accommodate the power from wind generators, and that subsea cables operate with an acceptable level of reliability. In doing so, policy makers ensure that offshore wind is smoothly integrated within the power system.

In closing, the offshore wind industry provides policy makers with enormous potential to meet a myriad of policy objectives. However, a consideration of the cost drivers, market structure, support scheme cost recovery method, social dimensions, and power market design is necessary to build up the capacity of the industry and lay the foundation for cost reductions and eventual phase out of subsidies. With the support of an effective policy framework, offshore wind has the potential to become a viable and mainstream renewable energy generation source in diverse markets around the world.

Index[1]

[1] Note: Page number followed by 'n' refers to notes.

© The Author(s) 2017
R. Poudineh et al., *Economics of Offshore Wind Power*,
https://doi.org/10.1007/978-3-319-66420-0

The manufacturer's authorised representative in the EU is Springer
Nature Customer Service Centre GmbH, Europaplatz 3, 69115 Heidelberg,
Germany. If you have any concerns regarding our products, please
contact ProductSafety@springernature.com

Printed and bound by CPI Group (UK) Ltd, Croydon, CR0 4YY
23/04/2026
02095601-0012